Introduction to Statistics with SPSS

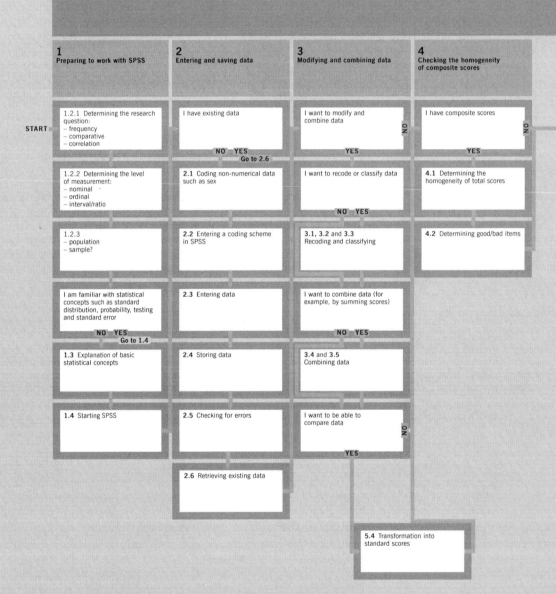

1
Preparing to work with SPSS

2
Entering and saving data

3
Modifying and combining data

4
Checking the homogeneity of composite scores

START

1.2.1 Determining the research question:
– frequency
– comparative
– correlation

I have existing data

NO   YES
Go to 2.6

I want to modify and combine data

NO

YES

I have composite scores

NO

YES

1.2.2 Determining the level of measurement:
– nominal
– ordinal
– interval/ratio

2.1 Coding non-numerical data such as sex

I want to recode or classify data

NO   YES

4.1 Determining the homogeneity of total scores

1.2.3
– population
– sample?

2.2 Entering a coding scheme in SPSS

3.1, 3.2 and **3.3**
Recoding and classifying

4.2 Determining good/bad items

I am familiar with statistical concepts such as standard distribution, probability, testing and standard error

NO   YES
Go to 1.4

2.3 Entering data

I want to combine data (for example, by summing scores)

NO   YES

1.3 Explanation of basic statistical concepts

2.4 Storing data

3.4 and **3.5**
Combining data

1.4 Starting SPSS

2.5 Checking for errors

I want to be able to compare data

NO

YES

2.6 Retrieving existing data

5.4 Transformation into standard scores

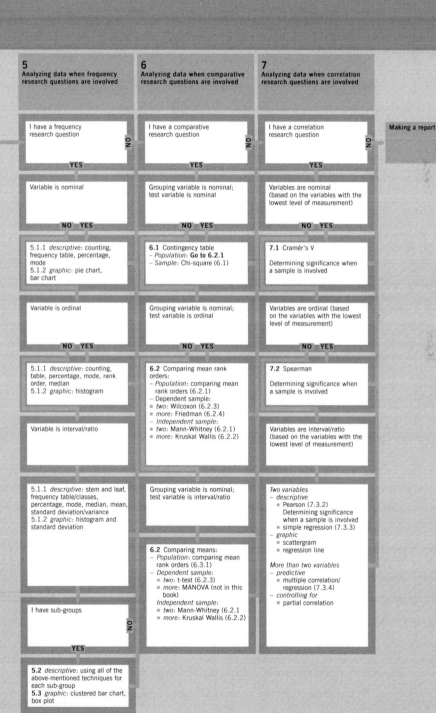

5
Analyzing data when frequency research questions are involved

6
Analyzing data when comparative research questions are involved

7
Analyzing data when correlation research questions are involved

Making a report

I have a frequency research question — NO

YES

I have a comparative research question — NO

YES

I have a correlation research question — NO

YES

Variable is nominal — NO YES

Grouping variable is nominal; test variable is nominal — NO YES

Variables are nominal (based on the variables with the lowest level of measurement) — NO YES

5.1.1 *descriptive*: counting, frequency table, percentage, mode
5.1.2 *graphic*: pie chart, bar chart

6.1 Contingency table
– *Population*: **Go to 6.2.1**
– *Sample*: Chi-square (6.1)

7.1 Cramér's V

Determining significance when a sample is involved

Variable is ordinal — NO YES

Grouping variable is nominal; test variable is ordinal — NO YES

Variables are ordinal (based on the variables with the lowest level of measurement) — NO YES

5.1.1 *descriptive*: counting, table, percentage, mode, rank order, median
5.1.2 *graphic*: histogram

6.2 Comparing mean rank orders:
– *Population*: comparing mean rank orders (6.2.1)
– Dependent sample:
  ▪ *two*: Wilcoxon (6.2.3)
  ▪ *more*: Friedman (6.2.4)
– *Independent sample*:
  ▪ *two*: Mann-Whitney (6.2.1)
  ▪ *more*: Kruskal Wallis (6.2.2)

7.2 Spearman

Determining significance when a sample is involved

Variable is interval/ratio

Variables are interval/ratio (based on the variables with the lowest level of measurement)

5.1.1 *descriptive*: stem and leaf, frequency table/classes, percentage, mode, median, mean, standard deviation/variance
5.1.2 *graphic*: histogram and standard deviation

Grouping variable is nominal; test variable is interval/ratio

Two variables
– *descriptive*
  ▪ Pearson (7.3.2)
    Determining significance when a sample is involved
  ▪ simple regression (7.3.3)
– *graphic*
  ▪ scattergram
  ▪ regression line

More than two variables
– *predictive*
  ▪ multiple correlation/regression (7.3.4)
– *controlling for*
  ▪ partial correlation

6.2 Comparing means:
– *Population*: comparing mean rank orders (6.3.1)
– *Dependent sample*:
  ▪ *two*: t-test (6.2.3)
  ▪ *more*: MANOVA (not in this book)
  Independent sample:
  ▪ *two*: Mann-Whitney (6.2.1)
  ▪ *more*: Kruskal Wallis (6.2.2)

I have sub-groups — NO

YES

5.2 *descriptive*: using all of the above-mentioned techniques for each sub-group
5.3 *graphic*: clustered bar chart, box plot

Introduction to Statistics with SPSS

Dr. D. B. Baarda, Dr. M. P. M. de Goede
en Dr. C. J. van Dijkum

Introduction to
Statistics with SPSS

*A guide to the processing, analysing and reporting
of (research) data*

Second, completely revised edition

Wolters-Noordhoff Groningen/Houten

Cover design: Total Identity, Amsterdam
Text Layout: Ebel Kuipers, Sappemeer

In the field of higher education, Wolters-Noordhoff BV publishes the imprints: Wolters-Noordhoff, Stenfert, Kroese, Martinus Nijhoff and Vespucci.

Any comments and remarks about these or other publications can be sent to: Wolters-Noordhoff BV, Higher Education Department, Antwoordnummer 13, 9700 VB Groningen, e-mail: info@wolters.nl

1 2 3 4 5 / 13 12 11 10 09

ISBN 90 207 32978
NUR 123

Preface

The extent to which a decision is accepted does not merely depend on the position in an organisation that the decision maker occupies, nor on a person's years of experience in the field, the degree of expertise in the area, powers of intuition or the fact that an individual may have a "good nose" for business. Such qualities are no longer sufficient, even if decisions taken on such bases are nevertheless good ones. After all, to be accepted, decisions not only have to be right, they must also be justifiable. Subjectivity is only one of the factors involved.

Sound decisions stem from conclusions based on outcomes derived from research. This involves the collection and analysis of data, as well as the reporting and dissemination of information. The ultimate foundation is provided by the data. Figures and statistics are objective, verifiable and, therefore, convincing.

Such views are currently becoming more widespread. Just open any given business magazine and you will find articles on this subject, all of them concerning the process of taking sound decisions. In this context, frequent use is made of such expressions as "Data Mining", "Business Intelligence" and "Statistical Analysis". What these notions have in common is the fact that they all involve data processing: the search for trends, problems, solutions or new (market) opportunities. Furthermore, they all promote the value of considering both subjective and objective factors whenever taking decisions.

With some pride, we can say that SPSS is by far the most frequently used statistical-analysis program. The *Introduction to Statistics with SPSS* constitutes the logical successor to *Statistiek met SPSS voor Windows*, a Dutch publication that appeared in 1999. As you will discover, SPSS has taken a further step forward in the areas involving ease of use and professionalism with the development of SPSS 12.0. SPSS can now be used to discover relationships and patterns, to detect causes, and to make well-founded decisions.

Good luck!
G.J. Hulzebos RIS (Registered Information Scientist)
Country Manager Products
SPSS Benelux B.V.

Foreword

What would we like to achieve in this book?

Once you have studied this book, you should:

1 *Have mastered the basic skills enabling you to work with SPSS on your own.*
This book provides practical instructions for processing and analysing research data using SPSS for Windows. SPSS stands for 'Statistical Products and Service Solutions', currently one of the most widely-used software packages for entering and statistically analysing data. The basic SPSS package contains programs for such commonly-used statistical analysis techniques as those involved in calculating a frequency distribution, cross tabulation or correlation coefficient.
In this *Introduction to Statistics with SPSS* for Windows, we hope that you will learn, through experience, to trust the software package. After a general introduction and familiarisation with the basic principles of SPSS, you will immediately be able to enter data into the computer and to analyze it. For this reason, we do not discuss the many extra features of SPSS in this book. The purpose of this book is to give you the basic skills so that you can work independently with SPSS. You will then inevitably learn how you can enhance these skills.
We invite the reader to begin by experimenting with the sample database concerning the relationship between wealth and happiness that we provide. This database contains survey data from 500 men and 500 women. We use it to investigate the usual practices involved in a research project.
• We begin with the entering of data (Chapter 2);
• we then explain how SPSS can be used to modify and to combine the data (Chapter 3);
• and in Chapter 4, we discuss how to establish the homogeneity of composite scores.

After this preparatory work, it is time for analysis. The methods used will depend on several factors including the specific nature of the inquiry's general research question. A well-designed general research question always involves one or more specific issues to which answers are sought by means of research. These specific research questions can further be distinguished into:

- frequency research questions, which ask how often or to what extent something occurs; techniques for describing such variables are investigated in Chapter 5;
- comparative research questions, which establish and test the similarities and differences between two or more groups concerning a given characteristic; they will be discussed in Chapter 6;
- correlation research questions, which establish and test the relationships between two characteristics; these will be discussed in Chapter 7.

2 *Know when to use which uni- or bivariate statistical method.*
Proper use of SPSS not only requires you to have some knowledge of SPSS, but also some expertise in the field of statistics. In the first chapter the basic concepts will be introduced, such as level of measurement, normal distribution, probability, significance, one and two-tailed tests, power and effect size. Chapters 5, 6 and 7 will reveal that the choice of statistical technique is not only dependent on the nature of the research question, but also on the level at which the relevant characteristic is measured and the issue of whether a population or a sample is involved.
Since this book only presents the statistical information that is strictly necessary to analyze data with SPSS using predominantly uni- and bivariate analytical methods, we will only deal with a few multivariate techniques of analysis.

3 *Know the conditions and SPSS procedure involved in a given technique.*
Using both explanation and example, we detail the essential factors in each technique without going too deeply into the underlying mathematics. We additionally indicate when a technique should and should not be used. Finally, there is an explanation of how to perform the technique using SPSS.

4 *Be able to read and to interpret SPSS output, and to issue reports based on it.*
The material will be further presented in terms of the following questions which you, the user, might pose:

- If the chosen method of statistical analysis has been successfully performed using SPSS, how do I read the computer output? What is precisely revealed by the results of the data analysis?
- How should I interpret the output? What does the output mean for my research question?
- How should I present my output in my research report?
- How do I write up my conclusions?

We will provide examples of SPSS output for each presented technique. On the basis of these examples, we will explain how the data should be read and what they mean. To illustrate this latter point, we additionally demonstrate how the output can be reported.

What are the instructional principles underlying this book?

A lot of consideration has been given to this book's instructional methodology.

Each chapter starts with an introduction that, in a general sense, provides a glimpse of that chapter's contents.

Each chapter ends with a list of key words in the order that they appear in the text. Additionally, the designated SPSS procedures are listed on a section-by-section basis, so that anyone who has an excellent or even reasonable command of SPSS for Windows need not work through the entire chapter again.

The main instructional premise is that you learn by doing. It is also important that you know what you are doing. For this reason, we also briefly explain the essential points of the technique to be used without going into detail. To gain insight into the research material, we advise you first to analyze the data graphically. First construct something like a scatter diagram to see what type of relationship is possibly involved. Depending on the research question, we discuss a few suitable graphic techniques that are available in SPSS.

The learning process is illustrated on the basis of a sample provided for practice entitled "Wealth and Happiness" or "Can money buy happiness?". Nearly all the problems encountered during the processing and analysis of data are explained by means of this practice example.

How can this book be used?

When you have conducted research and obtained a set of numerical values, you can use this book as a guide to the analysis of the data. It provides both information about statistics and instructions for using SPSS to analyze and interpret the figures. In fact, it is a book on statistics combined with an SPSS user manual. In principle, you do not need a separate statistics textbook and need not have any previous knowledge of the subject.

What do you need?

In writing this book, we have made use of SPSS for Windows, Version 12.0. This guide can, however, also be used with earlier versions of SPSS for Windows.

After you have worked through this book, you will be capable of quickly entering simple research data in a computer and adequately analysing them. You will likely also be able to perform statistical techniques that are not presented in this book but with which you are already familiar. The menus provide a great deal of information, along with relevant examples.

Where do I find the sample data files used in this book?
Statistical terms and methods are systematically presented and explained on the basis of concrete examples always related to an explicitly-stated, specific research question.
The sample research questions are developed in a step-by-step manner and illustrated in the boxes intended for that purpose. The data files on which we conduct the analyses can be found on the internet at http://www.basisboekstatistiekmetspss.wolters.nl.

Spring 2004

Ben Baarda
Martijn de Goede
Cor van Dijkum

Required hard- and software

This book explains how research data can be entered, prepared, analyzed and interpreted using SPSS for Windows and how a report can be written about the output.

You must be able to work with Windows (version 98, 2000, ME or XP) and must, of course, have installed SPSS. In this book, we make use of SPSS, Version 12.0.

Symbols and data files used

The following symbols are used in this book:

> Move the mouse pointer to the position indicated on the screen.

▶ Click the ▶button in SPSS.

⊤ Single-click the left mouse button.

⦿ Double-click the left mouse button.

The following data files are used in this book:

data1 raw, unprocessed data, entered according to the scheme of codes presented in Figure 2.1.

data2 recoded items rwlth1, rwlth2, rwlth3, rhap3, rhap4 and rhap5 have been entered here (see Section 3.1).

data3 the variables *ac* (age class; Section 3.3), *Twlth* and *Thap* (respective totals of the 5 wealth and 5 happiness items; see Section 3.4) have been entered here.

data4 the variable *mwlth* (the means of the 5 wealth items; see Section 3.5) and the items rwlth1_1 to rwlth5_1 (for the wealth items, the missing values are replaced by a series average; see Section 3.5) are entered here.

data5 the variable *Twlth* and *Thap* are here the respective sums of rwlth2, wlth4 and wlth5 and hap1, hap3 and hap4 (see Section 4.2).

family this is a data file containing the happiness scores of twenty men and their wives, along with the happiness scores from each of their oldest children.

The data files are stored on the website http://www.basisboekstatistiekmetspss.wolters.nl. It would be most helpful if these files were copied to a special 'data' folder/directory on the c-drive (hard drive) of the computer on which you are working.

Contents

1

How do I prepare myself to work with this *Introduction*?

For I don't care too much for money,
money can't buy me love

John Lennon & Paul McCartney

In conducting research, the choice of the appropriate statistical technique to analyze the collected data is an important link in the long chain of decisions that must be taken. Ultimately, the goal of collecting and analysing data is to answer the research question or questions contained in the inquiry.

To draw attention to the place of data analysis in the research cycle as a whole, the typical stages of research are listed below. Each stage in the research cycle is stated in the form of a question:

Prior knowledge
■ No specific prior knowledge is required.
■ Before explaining the uses of SPSS, we will first introduce the practice example: 'Wealth and happiness' (see Figure 1.1).

1 *What is the goal of my research and what is the type of the inquiry?*

2 *How will I acquire information (among other ways, by reviewing the literature)?*

3 *What type of research am I going to conduct?*

4 *What will my research design look like?*

5 *Will the entire population be involved in my research or will I select a sample?*

6 *Which data collection method am I going to use?*

7 *How will I prepare my data for analysis?*

8 *How will I analyze my data?*

9 *How will I report and evaluate my research?*

This book focuses on stages 7 and 8, along with a portion of 9: preparation, analysis and description of the research data analyzed using SPSS. Referring mostly to a study on the relationship between wealth and happiness (see Figure 1.1), we will investigate step-by-step:

- *How you should prepare research data for input (Section 1.1 and Chapter 2).*
- *How you can use SPSS to adjust and modify your data. Before you begin the analysis, you often have first to convert (recode) the values of certain data, or combine the values of data to produce new scores (Section 1.1 and Chapter 3), and subsequently check their reliability (Section 1.1 and Chapter 4).*

H1 How do I prepare myself to work with this *Introduction?*

- *How you choose the appropriate method of analysis. To choose the appropriate method of analysis, you first have to establish the nature of the research question (Section 1.2.1). Does it involve frequencies (Chapter 5), comparisons (Chapter 6) or correlations (Chapter 7). You then must determine the level of measurement for your data (Section 1.2.2). Finally, you have to decide if a sample or a population is involved in your research. You can then use the block diagram 'How do I analyze my data?' to ascertain which method of analysis best suits your research question. You will find the diagram on the inside front cover.*
- *How you use SPSS to conduct the analysis and how you should interpret the results. This we indicate for each statistical technique to be discussed. Additionally, we then examine how you can make a report about the results.*

This chapter further deals with a number of important statistical terms, such as normal distribution, significance, probability and standard error (Section 1.3). The final section of the chapter will explain how you can start using SPSS.

Befor explaining the uses od SPSS, we will first introduce the practice example: 'Wealth and happiness' (see Figure 1.1).

The questionnaire in Figure 1.1 was presented to a representative sample of 500 men and 500 women between the ages of 25 and 55 years old. These age limits were deliberately chosen. Many younger people are still studying and do not, therefore, have any fixed income. For people above the age of 55, there is often the issue of their (partial) withdrawal from the work force, as a result of which they enter another financial category. The data from this research is found on the website http://basisboekstatistiekmet spss.wolters.nl in the file entitled 'data1'.

21

Can money buy happiness?

A researcher would like to know if there is a correlation between wealth and happiness. He asks himself if money can buy happiness. His general research question can then be stated: 'Is there a positive correlation between the amount of wealth at an individual's disposal and the degree of happiness that the individual enjoys?' The concept "wealth" is defined by the researcher as follows: wealth is the quantity of financial means at an individual's disposal. This definition is deliberately broad. In this way, the research not only covers income and assets, but also other financial resources that individuals might have available to them. Happiness is defined as follows: happiness is the degree to which an individual is content about the life that he/she leads. The researcher has operationalized these two concepts in the form of a questionnaire. To measure both the concept of wealth and the concept of happiness, the researcher has devised five statements or items for each.

Wealth is measured by asking respsondents to agree or disagree with the following statements:

1	I own a car	yes/no
2	I own a home/apartment	yes/no
3	I own a DVD player	yes/no
4	I am covered by the Dutch national health plan	yes/no
5	I receive a rent subsidy	yes/no

Happiness is measured by asking respsondents to complete the following multiple choice items:

1 If I could live my life over again, I would…in the same way

☐   ☐   ☐   ☐   ☐
certainly not live it   not live it   to some extent (not) live it   live it   definitely live it

2 Most other people are … than I am.

☐   ☐   ☐   ☐   ☐
certainly not better off   not better off   to some extent (not) better off   better off   definitely better off

3 Things … as I would like them to be.

☐   ☐   ☐   ☐   ☐
are certainly not   are not   are to some extent (not)   are   are exactly

4 Life is …..

☐   ☐   ☐   ☐   ☐
certainly not difficult   not difficult   to some extent (not) difficult   difficult   extremely difficult

5 I … lonely.

☐   ☐   ☐   ☐   ☐
certainly do not feel   do not feel   do, to some extent (not), feel   feel   definitely feel

Since happiness is not only dependent on financial means but also on other factors, the researcher also asks about a number of easily-measured characteristics, such as the sex, age, educational level and marital/family status of the persons being surveyed.

Sex   ☐ male   ☐ female
Age (in years)   …
Marital/family statu   ☐ alleen
                ☐ met partner
                ☐ met partner en kinderen

Highest completed educational level
                ☐ Lower secondary education
                ☐ Upper secondary education
                ☐ Higher education

Figure 1.1 **Model research project 'Wealth and happiness'**

research questions

The researcher refined his general research question into a number of specific research questions:

1 How many people own cars, homes, or DVD players?
2 How many people are covered by the national health plan?
3 How many people receive a rent subsidy?
4 How happy are people about their lives?
5 How lonely do people feel?
6 Are there differences between men and women concerning:
   - ownership of a car, home or DVD player
   - national health plan coverage
   - reception of rent subsidies
7 Is there any difference in people's happiness about life when they are living:
   - with or without a partner
   - with or without children

8 Are there any differences in the loneliness felt by men and women?

9 Do men and women experience different degrees of happiness about the life that they are living?

10 Is there a correlation between contentment about life and age?

11 Is there any correlation between wealth and happiness?

1.1   How do I prepare the data for analysis?

Before you enter the data, you first have to make, for your own use, a summary of the variables involved in the research and the values that these could have. For example, the variable *sex* appears in our research project and has the possible values 'male' or 'female'. It is useful to convert these values to numerical form, for example '1' for male and '2' for female. You must, of course, appropriately note which values stand for what. You do this by means of a coding scheme or codebook. In Section 2.1, we explain how you can construct such a scheme.

coding scheme

When you have collected the research data and composed a coding scheme, you can enter the data in the computer. You do so by typing your data into the data editor, which appears on the screen when you start SPSS. It is additionally helpful if you immediately assign a name to the variables for which you are entering the scores. Before you begin the analysis of your data, you must check if values or codes are included that do not belong there. For sex, use is most frequently made of the codes '1 = male' and '2 = female'. The entries under sex in your database must, therefore, only be ones and twos. If, for example, a seven appears, you have probably made a mistake when entering the data. You need to correct it before continuing with your analysis.

data editor

Similarly, you have to check if the variables display sufficient variation. If there are very few men in your sample, it will be difficult for you to determine if a difference exists between men and women in, for example, the degree of their happiness. We investigate this further in Section 2.4 ("How do I check if I have made a mistake when entering my data?") and again in Chapter 5, when we deal with the calculation of frequencies.

variation

Sometimes, you have to process the data that you have entered. We are, for example, not only interested in a person's desire to live life over again, or the fact that someone may find life difficult, but above all the extent to which an individual agrees with such 'expressions of (un)happiness' . By adding up the scores from the separate items in the questionnaire, we can arrive at a total score for happiness. The sum, which is accomplished with the help of some SPSS commands, constitutes a new variable. You should have noticed that three of the five statements of (un)happiness have a negative formulation (for example: 'Life is difficult'). It would be helpful if all the scores for the happiness items 23 pointed in the same direction. In that way, a high score is always an indication of a high degree of happiness. Using a recode command, you can, once the data has been entered, adjust the values of all the variables by means of a single action, so that '5' becomes '1', etc. In Section 3.1, we explain how you can do that.

recode command

Sometimes you need to use more than one indicator to measure a concept. We have used five indicators to measure the concept 'wealth'. This translation of an abstract concept, such as wealth, into measurable, concrete characteristics, such as car ownership, is called operationalisation. If you use more than one indicator to measure a concept, such as 'wealth' or 'happiness' and then wish to combine these into one total score, you must invoke some criteria to check if this procedure is allowed. In short, the various operationalisations must all measure the same thing and therefore be homogeneous. With the aid of the homogeneity coefficient alpha (also known as the alpha reliability index), you can see if and, if so, the extent to which they all, in fact, measure the same characteristic. Using an item analysis, you can see if there are questions or items that have a negative effect on the homogeneity. If such is the case, you can decide not to include responses to such questionnaire elements in the total score for a concept. In Chapter 3 ('How do I modify and combine data'), we examine, among other things, how you derive a total score from the scores of the separate items. In Chapter 4, we use an example to provide a detailed answer to the question: How do I check the homogeneity of the collected scores?

operationalisation

item analysis

homogeneity

1.2 How do I analyze my data? A data-use key!
The answers to the following questions are important for the choice of method of statistical analysis.
1 Does the general research question involve issues of frequency (how often/to what extent), comparison or correlation? Or does it involve a combination of the above?
2 What is the level of measurement (nominal, ordinal or interval/ratio level) for the data that you have collected?
3 Is a sample or a population involved?

The block diagram 'How do I analyze my data' (to be found on the inside front cover of the book and as a separate insert) has been developed on the basis of these questions. The following sections explore the issue in greater detail.

1.2.1 What specific research questions are involved in my general research question?
The nature of the general research question will provide the basis for answering the question about which statistical technique to use. The general research question always contains one or more

specific research questions to which an answer must be given (Introduction).

In general, there are three types of specific research questions that can be distinguished in a research inquiry:

1 Questions concerning how often or to what extent something occurs (*frequency*). An example of this type is: 'To what extent are people in the Netherlands happy?' or 'What percentage of people possess cars?'
2 Questions concerning *a comparison*. Example: 'Are men happier than women?'
3 Questions concerning *a correlation*. Example: 'Is there any correlation between wealth and happiness?'

It is clear that research questions 1 to 5 (Introduction) can be characterized as frequency research questions. Research Question 1 is, for example, concerned with the number of people that own such consumer products as DVD players. Chapter 5 presents an example of the analysis used for data involving this type of query. Research questions 6 to 9 are comparative research questions. The analysis of data involving this type of inquiry is dealt with in Chapter 6. Questions 10 and 11 are correlation research questions. In Chapter 7, we examine examples of data analysis for research involving this type of question.

1.2.2   At what level can I measure my data?

Once you have determined the type of research question(s) involved in your research (see the first column in the block diagram 'How do I analyze my data'), you must then decide at which **level of measurement** the variable(s) has been measured. To do this, see the row below frequency, comparison or correlation in the above-mentioned block diagram.

For each research question, it is necessary to indicate which level of measurement is being used for the variable sex involved. In Section 1.1, the level of measurement for the variable sex in Research Question 9 (on the differences in the extent to which men and women feel themselves to be happy) is different from (and lower than) the level of measurement for the variables in Research Question 10, for example, on the correlation between age and happiness.

For the sex variable, there are only two categories, namely 'male' or 'female'. There is only a straightforward difference involved, no degree of more or less. A man is different from a woman but not more or less than her. The same holds true for marital status; people are married, living together or single. This type of possible answer concerns a **nominal level of measurement**. You can indicate how many men or women own a car, but not that someone is more 'male' or more 'female'.

25

Such gradations can indeed be made for data measured on the ordinal, interval or ratio levels.

ordinal level of measurement

Data on the ordinal level of measurement can definitely be described in terms of more or less, but this difference between categories cannot be expressed as a numerical value. For example, level of education is clearly an item describable as a comparison, some having more and others less. The upper high school level is higher than the lower one, but it is not possible to indicate how much higher.

interval and ratio levels of measurement

ratio level

interval level

At the interval and ratio levels of measurement, the difference between the categories conceivable as more or less is also expressible as a number. Temperature provides a good example of this type. The difference between 5 and 10 degrees Celsius is just as large as the difference between 45 and 50 degrees. In contrast to the interval level, the ratio level involves a natural zero-point, such as illustrated by weight or height. The interval level involves comparable intervals, but no natural zero-point. This distinction has consequences for the arithmetic calculation that may be used. With temperature, it is not possible to say that 20 degrees is twice as must as 10 degrees. With weight, it is conversely possible to say that 20 kilos is twice as heavy as 10 kilos. In the case of temperature, 0 degrees does not, after all, constitute the natural zero-point and, consequently, an interval level of measurement is involved. Despite this distinction, SPSS groups the interval and ratio levels of measurement under the heading of Scale. The two other levels of measurement are named Nominal and Ordinal.

Scale
Nominal
Ordinal

continuous variables

We additionally distinguish between continuous and discrete variables. Continuous variables allow you to present a line on which values form a row of connected points: a continuum. No matter how close to each other, any two given points are always separated by still (infinitely more) other possible values. Examples of continuous variables are a person's height, age and intelligence. Variables that can only have whole values are called discrete variables, such as the number of cars that someone can own or the number of children in a family.

discrete variables

1.2.3 Is a population or a sample involved?

There are two types of statistics: descriptive and inferential statistics. Descriptive statistics are used when research is conducted on a population. It is possible to speak of a population when all the units about which you wish to make statements are involved in your research. That is, when you conduct, for example, a survey of all the employees in a company in order to determine their level of job satisfaction.

descriptive statistics
population

26

sample

inductive or
inferential
statistics

To save costs, you could also solicit responses from just a portion (sample) of the employees that you select at random from the total employee pool. Of course, you still would like to make statements about the entire employee population. In such a case, use must be made of inductive or inferential statistics to allow you, on the basis of a particular case (a sample), to make general statements (about the population).

units

population study

sample study

Before you can begin the analysis of your data, you must therefore pose the question concerning which units (who or what) are the subject of your statements. When these are only the persons or items involved in your research, it is then possible to speak of a population study. If you would also like to make statements about persons or items not involved in the research but, so to speak, represented by the research units that you have selected, you are then undertaking a sample study. In Section 1.4, we briefly deal with a few statistical terms that are continually encountered when testing if the results in a sample are due to chance or if, within a certain margin of uncertainty, it is possible to make generalisations concerning the population from which the sample was drawn.

1.3 A few general statistical terms

The purpose of descriptive statistics is to present data in a clear and well-organized manner. If you have determined the job satisfaction of nearly a thousand employees, it makes little sense to present all this data. It is mostly, for example, compiled into a histogram (Section 5.1.2), or formulated as percentages or an average

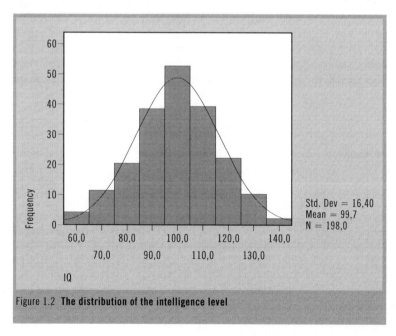

Std. Dev = 16,40
Mean = 99,7
N = 198,0

Figure 1.2 **The distribution of the intelligence level**

27

(Section 5.1.1). You describe your data in a reduced and therefore orderly form. If the data are reported in a graphic manner, the result is quite often the so-called 'normal distribution'. Figure 1.2 provides a (fictive) example of this.

It is the distribution of scores from an intelligence test administered to 198 employees of the 'Labour' company. This distribution closely approximates the form of a normal distribution. By way of comparison, the normal distribution is sketched. Such a distribution is also named the Gauss distribution or the Gauss curve. The characteristic feature of the normal distribution is its bell form. With SPSS, you can verify if the distribution of your data approaches the normal distribution (Section 5.1.2).

normal distribution

Gauss curve

Consider the possibility that 198 employees constitute a random sample from the total employee pool of the 'Labour' company (N = 2213). In such a case, use must be made of inductive or inferential statistics. The question then arises to what extent the mean IQ, found to be 99.7, is representative of the total employee population. In other words, what is the probability of the population having a mean IQ of 99.7 if you could include all of it in the research? This probability is, of course, not so great. For the mean that has been determined for the sample is somewhat dependent on the coincidental composition of the sample group. If we were to select another random sample and still another, etc., it is quite probable that the mean IQ could possibly be somewhat higher or lower. The values will likely deviate a little from each other, but probably not very much. SPSS can be used to calculate this so-called standard error (Section 5.1). The standard error indicates the extent to which the computed sample mean is a good estimate of the population mean. The larger the difference of IQ within the group and the smaller the sample size, the greater is the standard error. The standard error is therefore determined by sample size and sample homogeneity. On the basis of the standard error, you can, for example, state with at least a 95% certainty that the population mean lies between the sample mean minus twice the standard error on the one hand and the same sample mean plus twice the standard error on the other.

inductive or inferential statistics

standard error

sample size homogeneity

The notion of 'certainty' or probability plays an important role in inductive statistics. Even if you compare the means of two samples, there is some question about the probability that any detected difference in means is representative of the population.

probability

Consider the possibility that the sample of employees from the 'Labour' company consists of 99 women and 99 men. You find that the mean IQ of the women is 101.2 and that of the men is 98.2. Can you then state that the female employees of the 'Labour' company are on average more intelligent than the male employees? You can test if this difference is 'significant'. In Chapter 6, we explain how to do that when comparisons are involved and, in Chapter 7, how you do that when your research involves a corre-

significance

one or two- tailed
test
one-tailed test

lation. When is it then possible to speak of significance? The general rule is that a finding is significant if the probability of error (p value) is less than 5% or, with larger samples (> 1000) less than 1%. Often, SPSS also indicates if a one or two- tailed test was involved. You conduct a one-tailed test when you have formulated a hypothesis or expectation. If you have a theory on the basis of which you might expect that female employees are more intelligent than their male counterparts, you can then perform a one-tailed test. If you do not, however, have any idea about the possibility of a difference and certainly not about the direction of the difference, your test is then two-tailed.

The determination of significance is based on a few characteristics of the sample, in particular its size and homogeneity. The larger the sample, the smaller is the probability that any finding may be due to chance. The smaller the differences (i.e. range) of a given variable within a group (homogeneous groups), the smaller is the probability that the differences between groups involves a coincidence.

SPSS output also frequently contains the term degrees of freedom (df). The number of degrees of freedom indicate the extent to which scores can vary. If you know only one of two numbers (i.e. 36) and you know that the mean is 40, the other number must then be 44. This represents 1 degree of freedom. If you in fact know one figure, then you also know the other. For many tests, such as the t test (Section 6.3.1), the number of degrees of freedom is equal to the number of elements in the sample minus 1. For a contingency table (see Section 6.1) the number of degrees of freedom is equal to the number of rows minus 1, multiplied by the number of columns minus 1. For a 2x2 table, the number of degrees of freedom therefore equals 1. If the marginals (row and column totals) of a 2x2 contingency table are known as well as one of the cell frequencies, you can then calculate the other cell frequencies. Degrees of freedom are important when you wish to use a sample in order to estimate a population mean, for example. The significance of an observed difference or correlation in a sample depends on the number of degrees of freedom, which are in turn, except in the case of a contingency table, often dependent on sample size.

When you have established that there is, for example, a significant difference in mean IQ between men and women, you would often also like to know the extent to which the difference can be explained/predicted. To what degree can, for example, difference in mean IQ be attributed to sex? For this purpose there exists, depending on the manner in which the difference is tested, all sorts of measures that are often, however, not included in SPSS. We will explain how you can determine the effect size for each testing method that we discuss.

29

1.4   How does SPSS for Windows work?

Once SPSS has been properly installed on your computer, you should see the following icon on your screen:

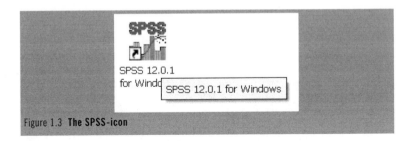

Figure 1.3 **The SPSS-icon**

You open SPSS by double clicking this icon (🕑). If you are unable to find the icon, you can always click the start button in the bottom left corner of your screen. You then use the mouse indicator to select "Programs". You can then see if SPSS appears on the program list displayed on screen. If such is the case, you can start SPSS by double clicking its entry in this list.

> SPSS icon 🕑

In SPSS Version 12.0, you then see the welcome screen as shown in Figure 1.4.

This welcome screen offers you various options. If you wish, for example, to enter *new data*, you have to click on "Type in data" and then "OK". An empty data matrix then appears. In earlier SPSS versions, the first screen that you saw was the empty data matrix. In Chapter 2 it is explained how you can enter data in the now open data editor.

data editor

You can also open a *previously-used, existing data file*. If it is a file that you have previously used, you will probably find it under the heading "Open an existing data source". You then click on the appropriate file followed by "OK", and the completed data matrix is displayed on screen.

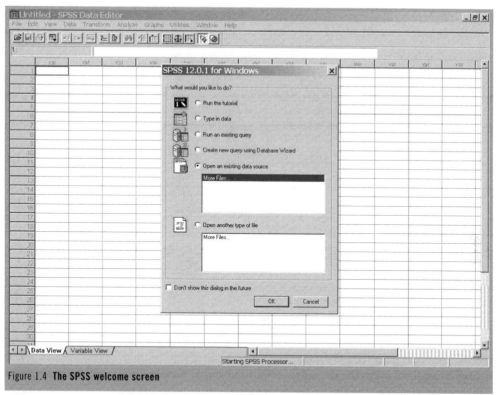

Figure 1.4 **The SPSS welcome screen**

When you wish to access a *still-unused, existing file*, such as the 'data1' file found on the http://www.basisboekstatistiekmet-spss.wolters.nl website, you click "Cancel". The small welcome screen then disappears, while the empty data matrix remains on screen. Now click "File" in the menu bar on the top left corner of the screen, followed by "Open" and "Data". You are then asked to indicate the name and location of the file. Typing in c:\data\data1, for example, and then clicking "Open", causes the sample 'data1' data file from the website to be opened. You will have needed first to have saved the data from the website to the 'data' directory on your c drive as a file named 'data1'.

> Cancel ⊙
> File ⊙
> Open
> Data ⊙
> type (for example) c:\data\data1
> Open ⊙

SPSS data matrix

The data in the 'data1' data file will then be entered in the SPSS data matrix. You can see this in Figure 1.5:

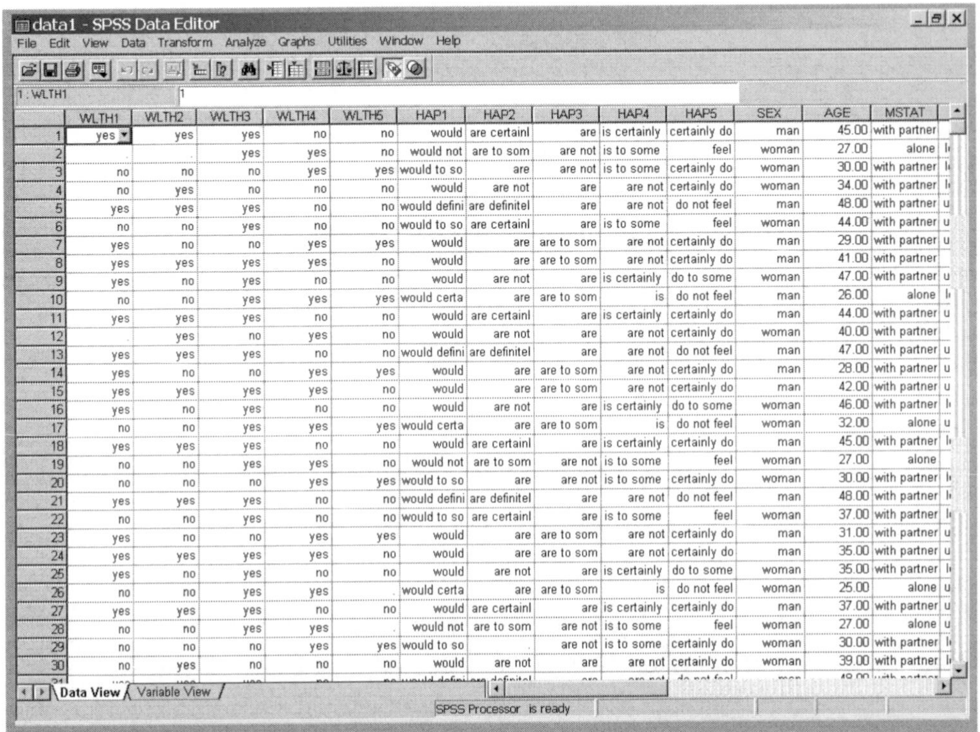

	WLTH1	WLTH2	WLTH3	WLTH4	WLTH5	HAP1	HAP2	HAP3	HAP4	HAP5	SEX	AGE	MSTAT
1	yes	yes	yes	no	no	would	are certainl	are	is certainly	certainly do	man	45.00	with partner
2	.	.	yes	yes	no	would not	are to som	are not	is to some	feel	woman	27.00	alone
3	no	no	no	yes	yes	would to so	are	are not	is to some	certainly do	woman	30.00	with partner
4	no	yes	no	no	no	would	are not	are	are not	certainly do	woman	34.00	with partner
5	yes	yes	yes	no	no	would defini	are definitel	are	are not	do not feel	man	48.00	with partner
6	no	no	yes	no	no	would to so	are certainl	are	is to some	feel	woman	44.00	with partner
7	yes	no	no	yes	yes	would	are	are to som	are not	certainly do	man	29.00	with partner
8	yes	yes	yes	yes	no	would	are	are to som	are not	certainly do	man	41.00	with partner
9	yes	no	yes	no	no	would	are not	are	is certainly	do to some	woman	47.00	with partner
10	no	no	yes	yes	yes	would certa	are	are to som	is	do not feel	man	26.00	alone
11	yes	yes	yes	no	no	would	are certainl	are	is certainly	certainly do	man	44.00	with partner
12	.	yes	no	yes	no	would	are not	are	are not	certainly do	woman	40.00	with partner
13	yes	yes	yes	no	no	would defini	are definitel	are	are not	do not feel	man	47.00	with partner
14	yes	no	no	yes	yes	would	are	are to som	are not	certainly do	man	28.00	with partner
15	yes	yes	yes	yes	no	would	are	are to som	are not	certainly do	man	42.00	with partner
16	yes	no	yes	no	no	would	are not	are	is certainly	do to some	woman	46.00	with partner
17	no	no	yes	yes	yes	would certa	are	are to som	is	do not feel	woman	32.00	alone
18	yes	yes	yes	no	no	would	are certainl	are	is certainly	certainly do	man	45.00	with partner
19	no	no	yes	yes	no	would not	are to som	are not	is to some	feel	woman	27.00	alone
20	no	no	no	yes	yes	would to so	are	are not	is to some	certainly do	woman	30.00	with partner
21	yes	yes	yes	no	no	would defini	are definitel	are	are not	do not feel	man	48.00	with partner
22	no	no	yes	no	no	would to so	are certainl	are	is to some	feel	woman	37.00	with partner
23	yes	no	no	yes	yes	would	are	are to som	are not	certainly do	man	31.00	with partner
24	yes	yes	yes	yes	no	would	are	are to som	are not	certainly do	man	35.00	with partner
25	yes	no	yes	no	no	would	are not	are	is certainly	do to some	woman	35.00	with partner
26	no	no	yes	yes	.	would certa	are	are to som	is	do not feel	woman	25.00	alone
27	yes	yes	yes	no	no	would	are certainl	are	is certainly	certainly do	man	37.00	with partner
28	no	no	yes	yes	.	would not	are to som	are not	is to some	feel	woman	27.00	alone
29	no	no	no	yes	yes	would to so	.	are not	is to some	certainly do	woman	30.00	with partner
30	no	yes	no	no	no	would	are not	are	are not	certainly do	woman	39.00	with partner

Figure 1.5 **The screen that appears ...**

Analyze

Graphs

You can now do a number of things. You can begin processing the data. You can also perform analyses by means of "Analyze", or compose graphs using "Graphs". In subsequent chapters, we will use concrete examples to elaborate these possibilities.

Questions that undoubtedly occur to you are: 'How can I shut the SPSS program off again?' and 'How can I obtain help when working with SPSS?'

You shut the SPSS program down by clicking on the small black cross in the top right corner of the screen. When the program is shutting down, you are asked if you wish to save any changes that you (may) have made to the file. If you indicate that you will *save* the modified file, you are requested to give the file a name. If you type in a new name, the modified file will be stored in addition to the unmodified version. If you leave the name unaltered, the modified file will be 'written' over the earlier version of the file.

When working with SPSS, you can receive help by clicking "Help" in the menu bar displayed on screen. In the roll-down menu that appears, you subsequently select "Topics". You then encounter the help menu, which is illustrated in Figure 1.6. If, for example, you no longer remember how you saved your data, you can type in 'save'. By clicking on "data files" and "Display", you obtain information about the procedure for saving your data. There are, additionally, help buttons for providing information about specific procedures involved in the various SPSS tasks themselves.

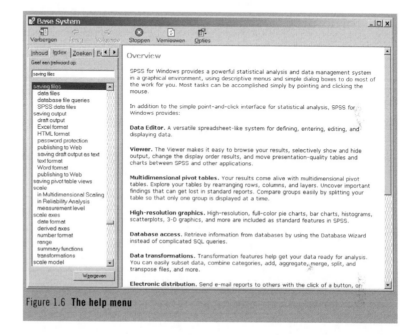

Figure 1.6 **The help menu**

Finally, we draw your attention to *all the various tutorials* included in SPSS. When you open SPSS, you see, among the series of options, that there is one entitled "Run tutorial" (see Figure 1.4). Clicking this option gains you access to a list of tutorials. In addition, SPSS has its own *website* containing all types of information about the program: www.spss.com.

Key words

Introduction
- research cycle
- research questions

1.1
- data preparation
- coding scheme
- codebook
- data editor
- variation
- recoding
- operationalisation
- item analysis
- homogeneity

1.2.1
- inquiry
- frequency research question
- comparative research question
- correlation research question

1.2.2
- level of measurement
- nominal level (in SPSS: Nominal)
- ordinal level (in SPSS: Ordinal)
- interval level (in SPSS: Scale)
- ratio level (in SPSS: Scale)
- continuous variables
- discrete variables

1.2.3
- population
- descriptive statistics
- sample
- inductive statistics
- inferential statistics
- units

1.3
- normal distribution
- Gauss curve
- standard error
- sample size
- homogeneity
- significance
- probability
- probability of error
- p value
- alpha (α)
- significant
- statistical testing
- one or two-tailed testing
- degrees of freedom (df)
- effect size

1.4
- data file
- data matrix
- data editor
- analyze
- graphs
- help function
- tutorials

34

2

How do I enter my data in the computer?

> *Money is the root of all evil.*

1 Timothy 6:10

■ We assume that:
- you are familiar
  with levels of
  measurement
  (1.3.2);
- you can start
  SPSS

*Before you enter your data, you must first clarify how the data
are to be read. It is best if all data are entered as numerical
values. The characteristic 'male' can, for example, be given the
value '1', which is then called the code for 'man'. You record
these values in a coding scheme. In this scheme, you can also
indicate what code to give to any missing data, such as any
unanswered questions (Section 2.1). In Section 2.2, we explain
how this scheme can be stored in the computer. You can then
enter the data: this is done by means of a data matrix in which
the variables are represented by the columns and the units by
the rows (Section 2.3). We explain in Section 2.4 how you can
save the entered data. Before you begin the analysis, you must
first check if you have made any typing errors and if there is
sufficient variation of data. It is possible that you may have
accidentally typed a '3' in the sex column, in which only ones
and twos may occur; it is important to correct this. It is, for
example, also possible that your data only includes responses
from men; there is then no dispersion in terms of sex and,
therefore, little sense in comparing the happiness experiences of
men and women. In the final section (2.6), we explain how you
can retrieve the stored data.*

2.1   How do I make a coding scheme or codebook?

Your data can best be entered numerically, that is as a set of figures. Even for such a variable as *sex*, it is, for example, more useful to enter a '1' in the case of a man and '2' in the case of a
woman. This does indeed mean that you will certainly have to remember that '1' stands for 'man' and '2' for 'woman'. It is prudent
coding scheme to record these codes in a coding scheme or codebook. In it, you
codebook indicate precisely and unambiguously which answer is assigned
to which code. The scheme must also be complete; all questions
and answers need to be included in it.

logbook It may also be helpful to maintain a logbook during your research. In this logbook, you keep a record of all decisions that you
took and all sorts of other things that are important for your research, such as your coding scheme. Every time that you start to
work on your research, you should open this logbook, which you
have made using a program such as Word. When you work with
SPSS, you can note important things directly in the logbook, such

36

as the given day that you stored a given item under a given name and in a given place. It may also be useful to indicate the changes that you make to a file.

variable

As can be seen in the coding scheme illustrated in Figure 2.1, every variable is given a name. To make it easy to distinguish the wealth variables from the happiness variables, we name them respectively WLTH1, etc. and HAP1, etc. Be careful to note that each question or item is a separate variable and, consequently, has a distinct name. Sometimes there can be more than one answer for a given question. In such cases, all *alternative answers must be treated as individual variables*. For example, you ask: 'Which means of transportation did you use to come here?' and the possible answers are: car, motorbike, moped, bike, walking, train, tram and bus. You now have to interpret each alternative, which is to make car, motorbike and the others, separate variables having the possible answers of yes or no.

codes

Besides the names of variables, the coding scheme also contains the codes for each of the indicated alternative answers, by which we mean the numerical value used for each answer. The coding scheme should also indicate what is to happen if any data are missing. If such occurs, the easiest procedure requires you to type in a blank space. It is standard for SPSS to recognize such a blank as a

missing value
system missing values

missing value and to identify it as one of the system missing values. SPSS then places a comma instead of a blank space in the data matrix. Do not, therefore, be surprised when you encounter a comma. You did not put it there, but it takes the place of a space. A disadvantage of handling missing data in this way is that no distinction is made between omitted data resulting from no answer being given and missing data due to some other reason; for example, the fact that the item was not applicable to the particular respondent in question. For this reason, it is often better to enter

user missing values

'user missing values' instead of the mentioned system missing values. Example: use a 9 if an answer is missing and an 8 if the item is not applicable. The latter would be the case for the question concerning rent subsidies when the respondent is young and living at home with his or her parents. If you would like the user missing values to be especially conspicuous, you could choose to designate them as 999 and 888 values. It is then important that these various missing values are specifically described in the codebook.

There is still another important tip for coding. As far as possible, *coding should occur in the direction of reading and writing*. For the items on happiness, we have awarded values from left to right. We award the code 1 to the 'definitely not' category of answer (far left) and the code 5 to the 'definitely' answer option (far right). This makes it easy to enter the data and prevents mistakes. With

recode

the help of the SPSS recode procedure, it is possible to recode the values of a variable after you have entered them. You can, for example, recode the negatively formulated happiness items so that a higher value is an indication of greater happiness. We deal with **37** the recode procedure in Section 3.1. Note: *When coding ordinal and interval/ratio data, such as age, it is better if they are not placed in classes (for example: 20-30 = young, 31-45 = middle-aged, > 45 = old) but entered in raw form.* Consider the case of several classes proving to be 'empty' (hence with a zero frequency), such instances can no longer be retrospectively altered. If you conversely enter the raw data for the same variables, you can then retrospectively implement all possible forms of classification.

Variable	Question	Answer code
WLTH1	I own a car	1=yes / 2=no
WLTH2	I own a home/apartment	1=yes / 2=no
WLTH3	I own a DVD player	1=yes / 2=no
WLTH4	I am covered by the National Health Service or some other form of socialized medicine	1=yes / 2=no
WLTH5	I receive a rent subsidy	1=yes / 2=no
HAP1	If I could live my life over again, I ... live it in the same way.	1 = would certainly not / 2 = would not / 3 = would to some extent (not) / 4 = would / 5 = would definitely
HAP2	Most other people ... better of than I am.	1 = are certainly not / 2 = are not / 3 = are to some extent (not) / 4 = are / 5 = are definitely
HAP3	Things ... as I would like them to be.	1 = are certainly not / 2 = are not / 3 = are to some extent (not) / 4 = are / = are exactly
HAP4	Life.... difficult	1 = is certainly not / 2 = is not / 3 = is to some extent (not) / 4 = is / 5 = is extremely
HAP5	I ... lonely.	1 = certainly do not feel / 2 = do not feel / 3 = do to some extent (not) feel / 4 = feel / 5 = definitely feel
SEX	Sex	1=man / 2=woman
AGE	Age	In years
MSTAT	Marital/family status	1=alone/ 2=with partner/ 3=with partner and children
EDU	Education	1=lower secondary/ 2=upper secondary/ 3=higher
Leave a space when information is missing		

Figure 2.1 **Example of a coding scheme**

The computer numbers the variables in the order in which you enter them. In many cases, this could be sufficient for your requirements. The output from such data may however be difficult to interpret, especially when your research involves several variables. It is therefore *advisable to give each of your variables a name.* When entering data in an older version of SPSS, it is standard for such names to consist of a *maximum eight characters*

labels

(letters and/or digits). You can also click the "labels" button to give variables a more elaborate name and even to provide the codes with a title. We advise you to do this when entering your data. It makes your output much more legible.

In Section 2.2, we will discuss how you can enter the coding scheme into SPSS by means of the "Variable View" tab in the bottom left corner (see figure 4).

2.2　How do I enter my coding scheme in the computer?

Variable View

The entering of the coding scheme, Figure 2.1, is done in "Variable View". Select therefore the tab having this name. You will

38

find it in the bottom left corner of the SPSS screen (Figure 1.4). In Figure 2.2, you can see an illustration of the screen that is subsequently displayed.

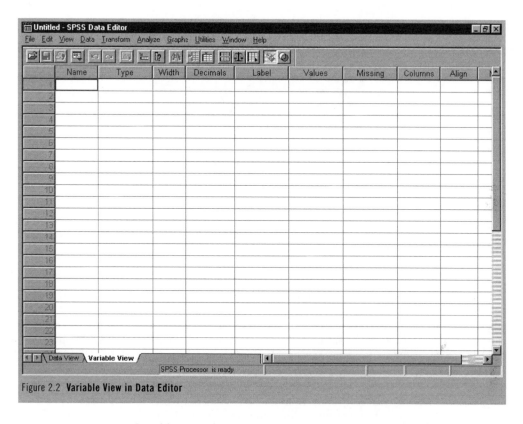

Figure 2.2 **Variable View in Data Editor**

In older versions of SPSS, the variable characteristics must be entered horizontally, along a single row. You can begin with the first row for the first variable. In the code book, you see that the variable is WLTH1. The name of this variable can be typed into the column entitled *Name* and may not consist of more than 8 symbols. The "Type" of the variable must then also be indicated. If you move the mouse indicator to the box on the right and press down the left mouse button, a number of choice options are displayed (see Figure 2.3).

Name
Type

Since SPSS is a calculator program, the software operates under the standard assumption that the variables are countable. You will therefore normally see a dot beside "Numeric" whenever you open the "Variable Type" window. In most cases, you will be able to tailor your work to suit this assumption and to leave the SPSS (default) preferences unchanged. In so doing, SPSS assumes that 8 spaces are sufficient (width) to allow you to indicate the number. It is therefore standard for SPSS to calculate using numbers that can consist of 8 digits. If you wish to work with more or, in

Numeric

default
width

39

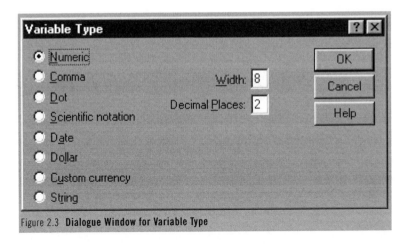

Figure 2.3 **Dialogue Window for Variable Type**

fact, fewer digits, you can indicate such in the "Width" field. SPSS also standardly allows two decimal places: two digits behind the decimal point. If you wish, you can also alter this to suit **Decimal Places** your purposes in the "Decimal Places" field. You can, for example, change it to 0 decimals when the question concerns the age of the subjects. If responses involve, however, such variables as income in euros or temperature in Celsius, it then makes sense to work with decimals.

In some cases, you might rather like, instead of numbers, to enter letters that may even form words. Such is the case when you wish **String** to type in names. You will then choose the "String" (meaning 'text') variable type. When you are later entering data, you must then also be careful to enter actual symbols and/or words, and you furthermore need to realize that statistical processes are difficult or impossible to perform on such input. Whenever possible, therefore, work with numerical variables and mark, preferably, the "Numeric" type.

Sometimes, it is important for you to provide an explanation for a variable; for example, the fact that HAP1 involves an individual's desire to live his/her life over again in the same manner. This is **Label** difficult to state in eight letters. You should then directly type your explanation in the "Label" column. In this way, you can prevent the possibility that you or other researchers will be uncertain about the meaning of a variable; for example, the fact that income is calculated in terms of net income per month in euros.

It is furthermore important that you indicate what a variable's possible values represent. For example, that insofar as sex is concerned, 1 is man and 2 woman. You do this by moving the mouse **Value** indicator to the right side of the "Value" field and then clicking the left mouse button to activate the dialogue window. See Figure 2.4.

40

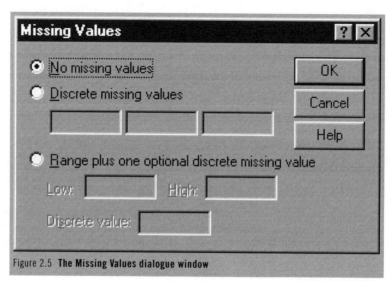

Figure 2.4 **The Value Labels dialogue window**

For the HAP1 variable, the codebook indicates that '1' is yes and '2' no. This is entered by typing '1' into the "Value" field and 'yes' into the "Value Label" box beneath it. You then click the "Add" button, which causes the first line in the "Value" list to appear. You then continue by typing in "Value" '2' and under that "Value Label" 'no', followed by the activation of the "Add" button to produce the second row. When you have entered all the possible values in this manner, including any *user missing values*, you can then activate the "OK" button. The list of possible values for the given variable has then been entered.

Value Label

user missing values

Next, the 'user missing values' can be defined. You should, however, realize that by indicating the value labels of user missing values, you are not entering the user missing values themselves. Move the indicator to the right side of the appropriate box and activate it with the left mouse button; you will then see the following dialogue window (see Figure 2.5).

Figure 2.5 **The Missing Values dialogue window**

41

You can select from a special list of user missing values. This can be a maximum of 3 complete (discrete) user missing values or a continuous range of such values, plus one distinct user value. For the WLTH5 (rent subsidy) variable, the following values can be entered: a 7 for 'no answer given', an 8 for 'an answer given but unclearly marked by the conductor of the survey' and a 9 for 'not applicable to this respondent'. In addition, the range 7–9 can be filled in. It is certainly important that, as far as possible, you use the same codes to indicate missing data when dealing with all your variables. Clicking "OK" causes the user missing value to be entered. If you wish to use a similar list for the following variable, the easiest procedure then involves (see Figure 2.2):

- moving the mouse indicator to the "Missing" box (but not on the right side of it!);
- pressing and releasing the left mouse button;
- going to the "Edit" menu and selecting "Copy";
- going next to the "Missing" field for the following variable;
- copying the list in that field by using the "Paste" command on the "Edit" menu.

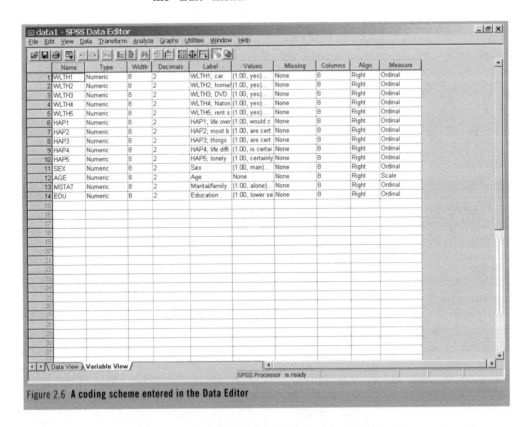

Figure 2.6 **A coding scheme entered in the Data Editor**

Columns

You can subsequently indicate in "Columns" the number of symbols that can be seen in the data-editor (see Figure 2.6); additionally, you can designate in "Align" (Right Justify) your alignment

42 **Align**

preference, which is more or less a cosmetic issue. Finally, the last column can be used to indicate the level of measurement ("Measure") at which the variables are being measured. Move the cursor again to the right side of the box, activate with the left mouse button, and then you can choose "Scale" (interval and ratio), "Ordinal" or "Nominal". An SPSS user has to know the level of measurement that an entered variable has otherwise you cannot select among the available statistical techniques.

Once you have provided all the above information for all the variables in the coding scheme, you will then see the list of variables appear on the screen, such as shown in Figure 2.6.

Review 2.1 is a summary of the steps that you must take to designate the variables to be entered.

Review 2.1

How do I indicate which variables are related to which values in my research?

> [tab in the bottom left corner of the Data Editor] Variable View ▼
>> > [in the first row of Variable View] cell in the Name column ▼
>> > type in the variable name
>> > cell in the Type column ▼
>> > right side ▼
>>> > leave standard Numeric setting
>>> > leave the standard setting (8) in the Width field unchanged
>>> > Decimal Places; type in the desired number of decimals
>>> > OK ▼
>> > cell in the Label column ▼
>> > type in explanation for variable
>> > cell in the Value column ▼
>> > right side ▼
>>> > move the mouse indicator to the Value field, and type in value, for example 1
>>> > move the mouse indicator to the Value Label field, and provide description
>>> > Add ▼
>>> > return the mouse indicator to Value type in the next value
>>> > to the Value Label following description of a Value
>>> > Add ▼
>>> > et cetera, just as long as required to enter all the values
>>> > OK ▼

43

> cell in the Missing column ▼
> right side ▼
> click one of the three options: No missing values;
  Discrete missing values; Range plus ... missing value
  [a dot appears in the circle beside the chosen
  option] ▼
> [beside 'Discrete ...' or 'Range plus ...'] fill in (User)
  Missing Values
> OK ▼
> cell in Columns ▼
> leave standard setting unchanged
> cell in Align ▼
> leave standard setting unchanged
> cell in Measure ▼
> define again your chosen level of measurement ▬▬

2.3 How do I enter my data in the computer?

2.3.1 What is a data matrix?

After making a coding scheme, you can enter data into the computer (see Figure 1.4 for the SPSS opening screen). In the left bottom corner of the SPSS screen, you will see the "Data View" tab. In "Data View", SPSS opens, by default, an empty data matrix. A data matrix is a co-ordinate system in which research units or respondents are represented by the rows (and are therefore listed down the side) and items or variables by the columns (and are hence listed across the top). Before you enter and analyze your research data, it is advisable that you first practice using a data matrix.

Data View
data matrix

To provide an opportunity for such practice, Figure 2.7 presents the data for the first ten people who participated in the 'Wealth and happiness' study. Use the coding scheme in Figure 2.1 to code the answers manually into the empty data matrix (Figure 2.8). To give you a little help beforehand: place a 1 (the respondent number) in the first cell of the first row; in the second cell of the same row place a 1 ('yes' in WLTH 1 is coded as '1'); the subsequent cells are filled by 1, 1, 2, 2, 3, 4, ... The coded data must then be entered in the computer. We will explain how this is done in Section 2.3.2. The data for the 1000 people who participated in the study can be found in the 'data1' file on the website.

2.3.2 Using Data View to enter data in the Data Editor

When you click the Data View tab, you are presented with an empty data matrix (Figure 2.9). Raw data, such as those provided in Figure 2.7, can now be entered in the matrix.

44

RESP	WLTH1	WLTH2	WLTH3	WLTH4	WLTH5	HAP1	HAP2	HAP3	HAP4	HAP5
1	yes	yes	yes	no	no	would	are certainly not	are	is certainly not	certainly do not feel
2	9	9	yes	yes	no	would not	are to some extent (not)	are not	is to some extent (not)	feel
3	no	no	no	yes	yes	would to some extent (not)	are	are not	is to some extent (not)	certainly do not feel
4	no	yes	no	no	no	would	are not	are	is not	certainly do not feel
5	yes	yes	yes	no	no	would definitely	are definitely	are	is not	do not feel
6	no	no	yes	no	no	would to some extent (not)	are certainly not	are	is to some extent (not)	feel
7	yes	no	no	yes	yes	would	are	to some extent are (not)	is not	certainly do not feel
8	yes	yes	yes	yes	no	would	better off	to some extent are (not)	is not	certainly do not feel
9	yes	no	yes	no	no	would	not better off	are	is certainly not	feel
10	no	no	yes	yes	yes	certainly not live it	better off	to some extent are (not)	is	do not feel

Figure 2.7 **The raw data from 10 persons (the rows) for 10 variables (the columns) in the 'Wealth and happiness' study**

RESP	WLTH1	WLTH2	WLTH3	WLTH4	WLTH5	HAP1	HAP2	HAP3	HAP4	HAP5
1	1	1	1	2	2	4				

Figure 2.8 **Empty data matrix for the 'Wealth and happiness' study'**

As shown, the variable names are each listed above the individual columns. You can now fill in the responses from each participant, beginning with the first variable, *wlth1*. To do this, you just type in the answer code for each specific answer; for example, enter the code '1' if the interviewee has answered 'yes' to the question if he/she owns a car (see coding scheme 2.1). Once you have done this for all variables and respondents, you should have a completed data matrix as shown in Figure 2.10.

45

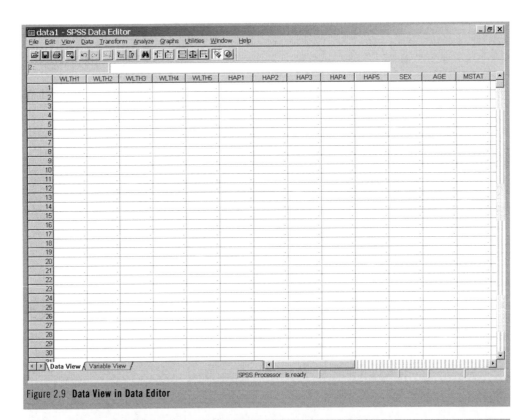

Figure 2.9 **Data View in Data Editor**

	WLTH1	WLTH2	WLTH3	WLTH4	WLTH5	HAP1	HAP2	HAP3	HAP4	HAP5	SEX	AGE	MSTAT
1	1.00	1.00	1.00	2.00	2.00	4.00	1.00	4.00	1.00	1.00	1.00	45.00	3.00
2			1.00	1.00	2.00	2.00	3.00	2.00	3.00	4.00	2.00	27.00	1.00
3	2.00	2.00	2.00	1.00	1.00	3.00	4.00	2.00	3.00	1.00	2.00	30.00	2.00
4	2.00	1.00	2.00	2.00	2.00	4.00	2.00	4.00	2.00	1.00	2.00	34.00	2.00
5	1.00	1.00	1.00	2.00	2.00	5.00	5.00	4.00	2.00	2.00	1.00	48.00	3.00
6	2.00	2.00	1.00	2.00	2.00	3.00	1.00	4.00	3.00	4.00	2.00	44.00	3.00
7	1.00	2.00	2.00	1.00	1.00	4.00	4.00	3.00	2.00	1.00	1.00	29.00	2.00
8	1.00	1.00	1.00	1.00	2.00	4.00	4.00	3.00	2.00	1.00	1.00	41.00	2.00
9	1.00	2.00	1.00	2.00	2.00	4.00	2.00	4.00	1.00	3.00	2.00	47.00	3.00
10	2.00	2.00	1.00	1.00	1.00	1.00	4.00	3.00	4.00	2.00	1.00	26.00	1.00
11	1.00	1.00	1.00	2.00	2.00	4.00	1.00	4.00	1.00	1.00	1.00	44.00	3.00
12		1.00	2.00	2.00	2.00	4.00	2.00	4.00	2.00	1.00	2.00	40.00	2.00
13	1.00	1.00	1.00	2.00	2.00	5.00	5.00	4.00	2.00	2.00	1.00	47.00	3.00
14	1.00	2.00	2.00	1.00	1.00	4.00	4.00	3.00	2.00	1.00	1.00	28.00	2.00
15	1.00	1.00	1.00	1.00	2.00	4.00	4.00	3.00	2.00	1.00	1.00	42.00	2.00
16	1.00	2.00	1.00	2.00	2.00	4.00	2.00	4.00	1.00	3.00	2.00	46.00	3.00
17	2.00	2.00	1.00	1.00	1.00	1.00	4.00	3.00	4.00	2.00	2.00	32.00	1.00
18	1.00	1.00	1.00	2.00	2.00	4.00	1.00	4.00	1.00	1.00	1.00	45.00	3.00
19	2.00	2.00	1.00	1.00	2.00	2.00	3.00	2.00	3.00	4.00	2.00	27.00	1.00
20	2.00	2.00	2.00	1.00	1.00	3.00	4.00	2.00	3.00	1.00	2.00	30.00	2.00
21	1.00	1.00	1.00	2.00	2.00	5.00	5.00	4.00	2.00	2.00	1.00	48.00	3.00
22	2.00	2.00	1.00	2.00	2.00	3.00	1.00	4.00	3.00	4.00	2.00	37.00	3.00
23	1.00	2.00	2.00	1.00	1.00	4.00	4.00	3.00	2.00	1.00	1.00	31.00	2.00
24	1.00	1.00	1.00	1.00	2.00	4.00	4.00	3.00	2.00	1.00	1.00	35.00	2.00
25	1.00	2.00	1.00	2.00	2.00	4.00	2.00	4.00	1.00	3.00	2.00	35.00	3.00
26	2.00	2.00	1.00	1.00		1.00	4.00	3.00	4.00	2.00	2.00	25.00	1.00
27	1.00	1.00	1.00	2.00	2.00	4.00	1.00	4.00	1.00	1.00	1.00	37.00	3.00
28	2.00	2.00	1.00	1.00		2.00	3.00	2.00	3.00	4.00	2.00	27.00	1.00
29	2.00	2.00	2.00	1.00	1.00	3.00		2.00	3.00	1.00	2.00	30.00	2.00
30	2.00	1.00	2.00	2.00	2.00	4.00	2.00	4.00	2.00	1.00	2.00	39.00	2.00

Figure 2.10 **Data in Data Editor**

H2 How do I enter my data in the computer?

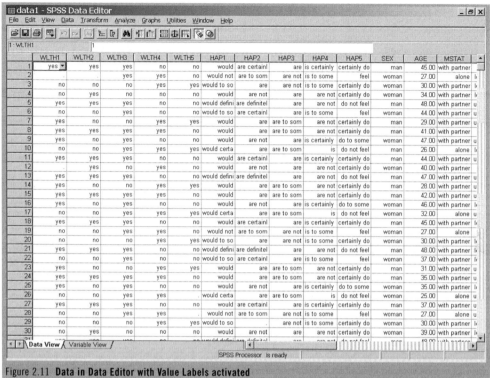

Figure 2.11 **Data in Data Editor with Value Labels activated**

View

Once you have done this correctly and placed a checkmark next to "Value Labels" in the "View" menu, the answer that each respondent has supplied is displayed, provided that you have, of course, properly entered the settings in "Value Labels". The result is illustrated in Figure 2.11.

Review 2.2

How do I enter data in the Data Editor?

> [tab in the bottom left corner of the Data Editor] Data View ▼
>> go to the first row, and click the first cell in the first column ▼
>> type in the code for the first respondent's answer to the first question, for example 1 ▼
>> go then to the second cell in the first row ▼
>> type in the code for the first respondent's answer to the second question ▼
>> type in the numerical codes for the first respondent's answers to question 3 ... to n
> then type the data in for respondents 2, 3, etc., each respondent being assigned a new row.

47

2.4 How do I save my entered data?

Undoubtedly, you will want to save the file in which you have entered your data. If you were, in fact, to end the SPSS session without saving the data (using save), the data would then be lost. If you wanted to conduct statistical analyses on these data in a subsequent SPSS session, you would have to enter them again. To store the data, you can save them by using the pop-down "File" menu:

> File ⓥ
> Save ⓥ

The first time that you wish to save your data, you have to indicate the location on a hard drive or diskette where you wish to save them, as well as the name under which they will be saved. For this reason, you must not click "Save" but "Save as".

> Save as ⓥ

If you wish to store the data on a diskette in the a: drive, in a file named 'data1' for example, you would then indicate the "file name" as follows:

> type: a:\data1

Next click "Save".
SPSS stores the file on your diskette under the name 'data1.sav'. SPSS automatically places the extension '.sav' after the name to indicate that the stored item is an SPSS data file. In this way, SPSS is able to recognize the file. SPSS can also readily identify other file types, such as the output files in which the results of an analysis are recorded. The ending that SPSS automatically adds to an output file is .spo.

If you would like to save the data in a 'data' directory on the c: hard drive, you can do that in the following manner:

> type: c:\data\data01
> Save ⓥ

It goes without saying that there must indeed be a 'data' directory on the c: drive.

It is useful if you choose a name for your data file that immediately makes it apparent what type of data file it is. As indicated, SPSS can also identify other files. It is also a good idea to give the file a number, which increases by one for every additional version of the file: hence 'data01', data02', and so on. If you make a mistake, you can always resort to previous versions of your data file without, consequently, having lost everything. It is advisable to save the file immediately at the beginning by means of the "Save as" command. Your file immediately receives the name that you want to give it and is no longer called "New Data". Every now and then you can save your data by clicking on the save icon: the diskette on the menu bar.

> save icon to save data (diskette with arrow) ⓥ

Margin notes:
save
File
Save
Save as
.sav
SPSS data file
output file
.spo

It is furthermore recommended that you save your data in at least two ways: on the hard drive and on a diskette. If something goes wrong, you will always have an extra copy. Your logbook should contain an indication of where and under which names the data are stored.

Review 2.3
How do I save the entered data?

> [in menu bar] File ⊙
> > Save as....
> > > place the mouse indicator in the File name field ⊙
> > > type a name, for example: a:\data1
> > > Save ⊙
> > or
> > save icon (diskette) ⊙

2.5   How do I check if I have made any mistakes when entering my data?

coding and/or typing errors

After entering your data, you have to check if any coding and/or typing errors have been made. If you should, for example, encounter a 3 among the entries for the *sex* variable, an error has then clearly been made. After all, this variable can only have two codes: 1 for man and 2 for woman. By printing out a frequency distribution for all variables, you can quickly see if any mistakes were made. This is done as follows. We assume that the SPSS main menu is currently being displayed to you. For safety's sake, first save your data file (once more). Move the mouse indicator to the menu bar so that it indicates "Analyze" (see Figure 2.12).

frequency distribution

Analyse

Descriptive Statistics Frequencies

In this pop-down menu, you first select "Descriptive Statistics" and then "Frequencies":

> Analyze ⊙
> > Descriptive Statistics
> > > Frequencies... ⊙

You now see a table containing all the variables. Move the mouse indicator to the first variable. You then press the left mouse button and drag the mouse indicator to the last variable. You can also place the mouse indicator on the first variable and then depress the Shift key and hold it down while you select the other variables. If you have done it properly, all variables will be selected. Move next to the right side of the box under the heading "Variables" by clicking the arrow button between the two windows. Clicking "OK" causes the procedure to be executed.

49

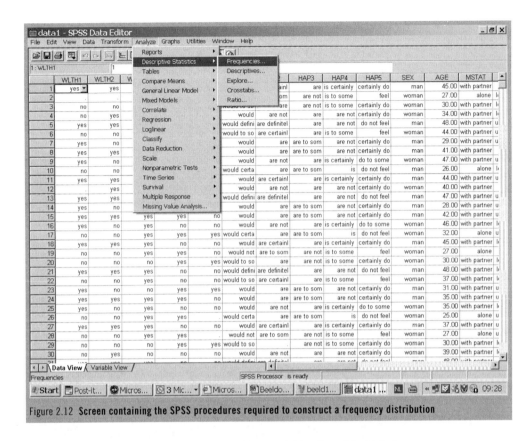

Figure 2.12 **Screen containing the SPSS procedures required to construct a frequency distribution**

> first variable ⊙ and hold it down
> last variable ⊙
> ▶ ⊙
> OK ⊙

The program will now be run and the output will appear on the screen. If everything has been done properly, you will have a result similar to that in Figure 2.13. Note: this is another type of screen, the output screen. If the output screen has its maximum size, it appears as if the data- input screen has disappeared, but this is not so. If you click "Window" in the top menu bar, you will see that the data editor screen is still there. In addition to the data editor and the output screen, SPSS can also display a chart screen. This becomes visible when you use SPSS to construct a graphic, for example. We will return to this subject in Section 5.1.2.

output screen

data editor screen
chart screen

No coding or typing errors were made in coding Wlth1 and Wlth2. Only the values 1 (=yes) and 2 (=no) occur in the data for both variables. Now compare the frequency distribution of these variables with the corresponding data in Figure 2.10. In Section 5.1, we will provide more details about the reading and interpreting of computer output.

50

wealth 1: car

		Frequency	Percent	Valid Percent	Cumulative Percent
Valid	Yes	541	54.1	55.7	55.7
	No	430	43.0	44.3	100.0
	Total	971	97.1	100.0	
Missing:	System	29	2.9		
Total		1000	100.0		

wealth 2: owner-occupied house/apartment

		Frequency	Percent	Valid Percent	Cumulative Percent
Valid	Yes	476	47.6	48.5	48.5
	No	506	50.6	51.5	100.0
	Total	971	97.1	100.0	
Missing:	System	18	1.8		
Total		1000	100.0		

Figure 2.13 **SPSS output screen displaying a frequency distribution for two wealth variables**

Check if there are any values that do not belong in your data. In the case of the *sex* variable, for example, only a '1' or a "2" may appear and no other number. Consider the case that a '3' should occur, you can then search the data matrix to find where the '3' appears and change this to the correct value.

variation

Furthermore, it is advisable to examine if the variables display any variation. A characteristic of a variable is the fact that it possesses a certain range of values, such as '1' (man) and '2' (woman). If only ones should occur, that is, only men, the variable is then actually uninteresting insofar as any further analysis is concerned. It is then, of course, not possible to make any comparison between men and women.

In Review 2.4 we will summarize the procedures that you need to perform to check if any (coding) mistakes have been made.

Review 2.4

How do I check if any typing errors have been made when entering data?

> [in menu bar] Analyze ▼
> > Descriptive Statistics
> > > Frequencies... ▼
> > > select all variables by using a drag procedure or the Shift key
> > > ▶ ▼ [window entitled Variable]
> > > OK ▼

2.6   How do I retrieve my entered data?

If you wish to perform given analyses on the data that you have previously entered, you can retrieve your data by means of the 51

Open

opening menu (see Figure 1.4), in which a list of the most recently used files is displayed. You can also make use of the "Open" command. The latter is found on the pop-down menu that appears when you click the "File" command on the menu bar. When you indicate that you would like to open a data file, SPSS searches the data files in the SPSS directory. If the data is located on a diskette (most commonly in the a: drive), you then have to indicate that fact. For the file name, you then type 'a:\data01'. Click next on "Open" in order to open the file. You can now add data or conduct analyses. If you add more data, store the enlarged file under a new name, for example, 'data02', and note this fact in your logbook. If you no longer know where you have stored your data file, take a look in your logbook.

Review 2.5
How do I retrieve my entered data?

> in the opening menu, first click the appropriate file and then "OK"
or
> [in menu bar] File ⊙
  > Open ⊙
    > Data ⊙
      > fill in file name
      > Open ⊙

Key words

2.1
■ code book, coding scheme
■ logbook
■ variable
■ coding
■ missing values
■ system missing values
■ recoding
■ variable label

2.2
■ variable view
■ variable name
■ variable type
■ numeric
■ default value
■ number of spaces (width)
■ decimal places
■ string
■ variable label (label)
■ description of values
■ column width
■ alignment
■ level of measurement

2.3.1
■ data view
■ data matrix

2.3.2
■ data entry
■ data view

2.4
■ file
■ save
■ save as
■ data file (.sav)
■ output file (.spo)

2.5
■ coding and/or typing errors
■ frequency distribution (analyze > descriptive statistics > frequencies)
■ output screen
■ data-editor screen
■ chart screen
■ variation

2.6
■ opening a file (file > open > data)

52

3

How do I modify or combine data?

Making money ain't nothing exciting to me.
You might be able to buy a little better booze
than the wino at the corner.
But you get sick like the next cat,
and when you die you're just as graveyard dead.

Louis Armstrong

1	2	3	4	5	6	7
How do I prepare myself to work with this Introduction?	How do I enter my data in the computer?	How do I modify or combine data?	How do I verify the homogeneity of the composite scores?	How do I analyse my data when a frequency re-search question is involved?	How do I analyse my data when a comparative re-search question is involved?	my data when a correlation re-search question is

3.1	3.2	3.3	3.4	3.5
What must I do to recode data?	How do I save the recoded data?	How can I divide a variable into classes?	How can I combine data?	How can I take missing values into account when combining data?

In the section of the questionnaire dealing with happiness, some

of the statements have a positive formulation; others are

formulated in a negative manner. This means that happiness is

indicated by a high score in some items and, conversely, a low

score in others. The state of affairs is unruly and confusing. The

situation can be remedied by reversing the order of the values

derived from the negatively formulated items. In SPSS, this

reversal of sequence is executed by means of a recoding

procedure (Section 3.1). In Section 3.2, we explain the procedure

by which you can save your recoded data. You can, however,

process the values of the entered data in various ways.

Sometimes, as in the case of age, it is useful to construct classes

or categories. In Section 3.3 it is explained how you use SPSS to

perform such a classification. SPSS can also be used to combine

data; you can calculate a total score for happiness, for example

(See Section 3.4). In performing such an operation, it is certainly

important to take missing data into account (Section 3.5).

Prior knowledge
- We assume that:
- you can start SPSS for Windows (1.4);
- you know what a coding scheme is (2.1);
- you know what a data matrix is (2.1);
- you can use SPSS to make a frequency printout (2.5)
- you can retrieve a data file (2.6).

3.1   How to recode data?

In the 'Wealth and happiness' study, respondents are presented with a number of questions. Those about wealth involve the ownership of a car, house/apartment or DVD player; type of health insurance (whether or not they are covered by the national health plan); and eligibility for a rental subsidy. In each of these five questions, the code '1' denotes the answer 'yes' and the code '2' the answer 'no'. You will have noticed that an affirmative answer to the first three items indicates that the respondent enjoys higher levels of well-being and income than an individual who answers in the negative. For the statements concerning "national health-plan coverage" and the reception of a "rent subsidy", an affirmative response indicates, conversely a lower level of well- being or income. Such a scoring system is inconsistent. You would expect higher well-being and income to correspond to a higher score or code, and lower well being to a lower score. The scores from questions 1, 2 and 3 can, however, be inverted with the aid of SPSS.

You might ask why the codes for these wealth items are not reversed (in the coding scheme) at the time when the variables are being defined. This is also possible. It is, however, easier for the person entering the information if similar questions all have the same codes corresponding to the same answers (hence always code '1' for 'yes' and code '2' for 'no'). If the coding is constantly changed in a long questionnaire, the chance of making an error when entering the information is great. For this reason, it is much simpler, once all the information has been entered, to employ a simple SPSS command in order to institute a retrospective change or recoding of the designated variable for all the respondents at the same time. This reversing of the order or recoding is done by means of the "Recode" command. You find this command in the roll-down menu displayed when you click the "Transform" button on the menu bar (see Figure 3.1). Assuming that you have first retrieved the file containing the data (in our case 'data1.sav'), the procedure is as follows:

reversing of the order
recoding
Recode
Transform

> Transform ▼
>> Recode

Once you indicate that you want to use "Recode", the programme first asks you if you would like to store the new information under the same name ("Into Same Variables...") or if you wish to make a new variable ("Into Different Variables..."). To avoid confusion with the original data, it is useful to choose a new name. Hence:

Into Same Variables
Into Different Variables

> Into Different Variables... ▼

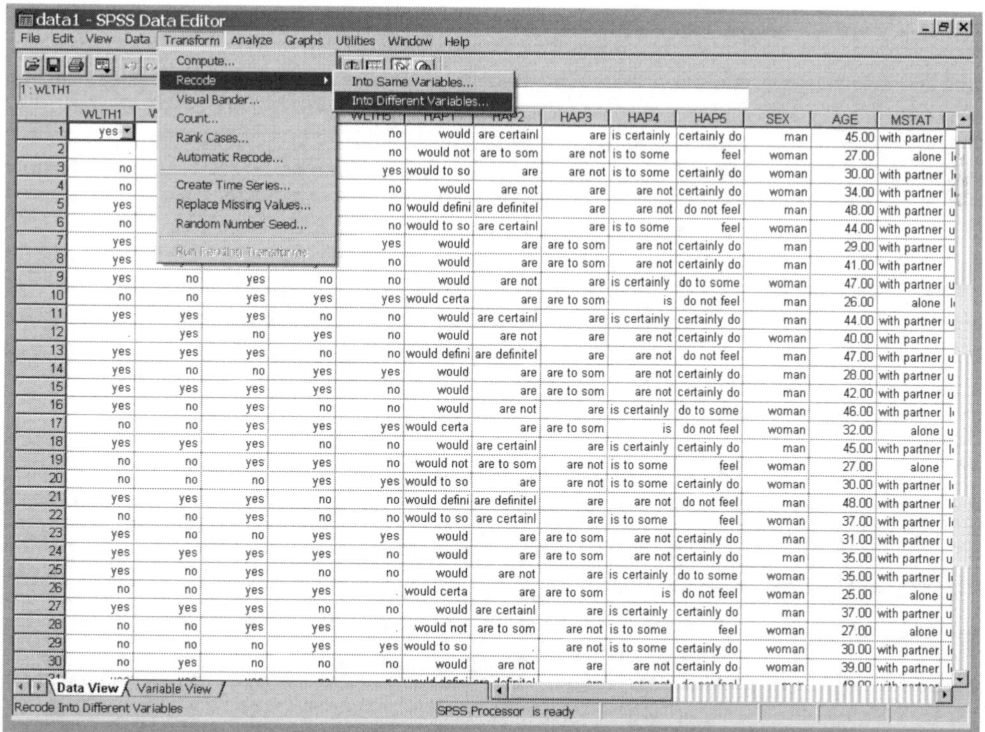

Figure 3.1 The recode procedure

You add, for example, an 'R' for recoded so that WLTH1 becomes RWLTH1. In the window for the "Recode" procedure, you first indicate which variables you want to recode. You do this by selecting them and moving them to the window entitled "Numeric Variable → Output Variable" by clicking the ▶.

Mark the variables that you wish to recode, for example:
> WLTH1 WLTH2 WLTH3 ⊙
> ▶ [move to the Numeric Variable → Output Variable field] ⊙

The programme then asks for the new name to be given to the designated variables. Move the mouse indicator to the "Name" field in the "Output Variable" box and click it once. Then type the new name in the field, for example, RWLTH1, and confirm it by clicking "Change" (see Figure 3.2).
> WLTH1 ⊙
> Output Variable: Name ⊙
> Type the new name: RWLTH1
> Change ⊙

Old and New Values
Add

You will see that you can immediately give the recoded variable a label. The label can be typed into the "Label" field. You must then indicate the manner in which these variables will be recoded by means of the "Old and New Values" procedure. In "Old Value", indicate the existing value, for example '1'. Next, type '2' in "New Value". Do not forget to click "Add". The change then appears in the table. Do the same for the change of '2' to '1'. Again, do not forget to click "Add".
> Old and New Values ⊙
> Old Value: type the existing value '1'
> New Value: type the new value '2'
> Add ⊙
> Old Value: type the existing value '2'
> New Value: type the new value '1'
> Add ⊙

You can use the same recoding function to *recode several variables* at the same time. Of course, each of the selected variables must be given a 'new' name. In this example, such is the case for the variables WLTH2 and WLTH3. Click "Continue" followed by "OK". If everything has been done properly, the new variables will appear in the data matrix at the end of the row, along with their (recoded) values:
> Continue ⊙
> OK ⊙

H3 How do I modify or combine data?

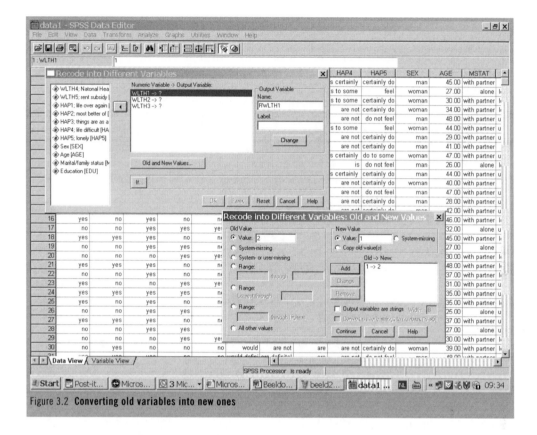

Figure 3.2 **Converting old variables into new ones**

Review 3.1

How can I recode variables?

> Transform ⓥ

  > Recode

    > Into Different Variables... ⓥ

      > mark the variables to be recoded, e.g. VAR1, VAR2, ...VARn

      > ▶ [move to the Numeric Variable → Output Variable field] ⓥ

      > VAR1 ⓥ

      > Output Variable Name ⓥ

      > type the new name: RVAR1

      > Change ⓥ

      > Old and New Values... ⓥ

      > Old Value: type the existing value, e.g. 1

      > New Value: type the new value, e.g. 2

      > Add ⓥ

      > Old Value: type the existing value, e.g. 2

      > New Value: type the new value, e.g. 1

      > Add ⓥ

       and so on, until you have processed all the variables that you would like to recode.

      > Continue ⓥ

    > OK ⓥ

57

As was the case for the wealth variables, the variables for happiness are not coded in the same direction either. Happiness variables 1 and 3 have a positive formulation, while happiness variables 2, 4 and 5 are formulated in a negative way. It would be sensible to recode the last three variables so that, similar to the codes for the wealth statements, a high score indicates a higher level of happiness. This means that the value '5' must be converted to '1', '4' to '2', '3' remains '3', '2' becomes '4' and '5' is changed to '1'. Make sure that the value '3' is unaltered; type a '3' in both the "Old Value" and "New Value" fields. In this way, you indicate that the value remains the same, as this does not happen automatically.

3.2   How do I save the recoded data?

The changes that you have now made remain in effect as long as you continue working in the current SPSS session. It will be necessary to save the recoded values for the variables RWLTH1, RWLTH2 and RWLTH3, as well as RHAP2, RHAP4 and RHAP5. This occurs in the usual way. Select "File" in the menu bar followed by "Save as". You then type the new name in the field of your data file containing the recorded variables:

Save as

> File ⊙
> > Save as ⊙
> > type: a:\data2.sav
> > Save ⊙

This procedure names the file containing the recoded values as 'data2' and writes it to the diskette in the a: drive. You should also make a note in your logbook indicating that the recoded values have been stored in 'a:\data2'. The next time that you retrieve the 'data2' file, you will not have to recode the responses to the designated variables all over again. Do not forget that these variables have been given values that are different from those of the originals. This must also be noted in your logbook. It is even more helpful to give each of the new values a new label, which you can do by means of the "Value Label" procedure (Section 2.2).

3.3   How can I divide a variable into classes?

categories
classes

By using the "Recode" procedure, the values for such variables as age and weight can be grouped into categories or classes. In general, it is undesirable to subdivide variables into classes. After all, the scale is then somewhat fragmented and information resultantly lost. It is certainly imprudent to use classes when entering the information for these variables. Suppose that we, when entering the data for the *age* variable, had divided them into the following groups:

58

≤ 20	code 1
21-25	code 2
26-40	code 3
41-55	code 4
56-65	code 5
≥ 66	code 6

As a result of the above method, we would be working with much less differentiated data than was, in fact, possible. A person age 41 would, after all, be placed in the same age class as someone who is 55. Furthermore, any age data entered according to this classification can no longer be converted back to their original ungrouped state. *It is better for you to enter the raw data and then, if there is cause to do so and with the help of SPSS, divide them into classes.* Once you have entered the raw data, you can revise any classification that you may have implemented if, for example, one or more classes prove to be empty.

The formulation of a system of classes is also a task that can be executed by means of the "Recode" command. To provide an example, we will recode the *age* variable. Suppose that you wish to make a distinction between older and younger respondents. It is presumed that older individuals have, on average, more money than younger people.

As indicated in Section 3.1, you are required to execute the following procedures:
> Transform ⊙
  > Recode
    > Into Different Variables... ⊙

You next darken the *age* variable in the variable list and place it in the "Numeric Variable → Output Variable" field (see Figure 3.3).
      > AGE ⊙
      > ▶ [move to the Numeric Variable → Output Variable field] ⊙

To distinguish the *age class* variable from the variable of *age (in years)*, you can give the former another name, for example AC. The C then indicates that a system of classification has been used.
      > Output Variable Name ⊙
      > type: AC

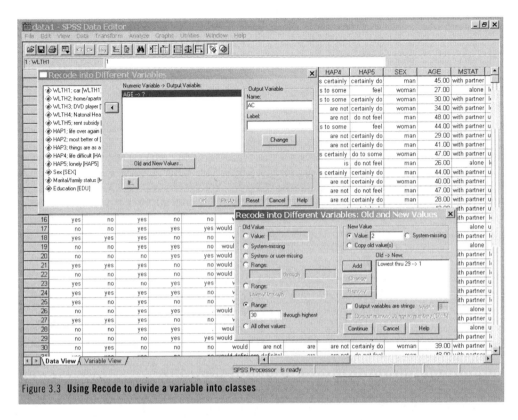

Figure 3.3 **Using Recode to divide a variable into classes**

Next, you have to indicate the manner in which the data are to be divided into classes by means of the "Old and New Values" procedure. We choose to distinguish 'younger than 30' from '30 years and older'. You have then no need to know the lowest score in the *age* variable to make this classification. It involves two classes: the ages ranging from the "lowest score through 29" and ages from "30 through the *highest* score". The procedure is as follows:

> Old and New Values... ⊙
> Old Value: Range: Lowest through ... ⊙

Range: Lowest through ...

A black dot appears in the selection circle in front of this command. It indicates that your choice has been processed and that you can continue to define your classes:

> type: 29
> New Value: Value ⊙
> type: 1

You must then confirm your input by clicking "Add":

> Add ⊙

H3 How do I modify or combine data?

You indicate the boundaries of the second class in the same manner:

Range: ... through
highest

> Old Value: Range: ... through highest ⊙
> type: 30
> New Value: Value ⊙
> type: 2
> Add ⊙

The change is executed by clicking "Continue" followed by "Change". The procedure is concluded by means of "OK":

> Continue ⊙
> Change ⊙
> OK ⊙

In Review 3.2, we once more summarize the procedures that you use to construct classes within a variable.

Review 3.2

How do I construct classes within a variable?

> Transform ⊙
> > Recode
> > > Into Different Variables... ⊙
> > > VAR1 ⊙
> > > ▶ [move to the Numeric Variable → Output Variable field] ⊙
> > > Output Variable Name: type cVAR1
> > > Old and New Values... ⊙
> > > Old Value: Range: Lowest through ... - type: [numeric code]
> > > New Value: Value: > type: [numeric code]
> > > Add ⊙
> > > Old Value: Range: ... through highest - type: [numeric code]
> > > New Value: Value: > type: [numeric code]
> > > Add ⊙
> > > Continue ⊙
> > Change ⊙
> OK ⊙

3.4   How can I combine data?

new variable

In some cases, it is useful to combine data in order to compose a new variable. To illustrate this point, we will calculate a total score by combining the responses to the five wealth statements. The task requires these responses to be (re)coded, so they are all scaled in the same direction. The scales on which responses to items 1 to 3 are scored must, therefore, be inverted, and the re-coded data for these variables stored in the file 'data2.sav'. Fur-   61 thermore, it goes without saying that all five items (variables WLTH1 through WLTH5) must more or less measure the same thing. Chapter 4 provides an explanation of how you can verify this. To combine responses to constitute a total score for wealth,

Compute
you use the "Compute" command on the "Transform" menu.

> Transform ⊙
> > Compute... ⊙

Target Variable
When you click "Compute", you open a screen in which, among other things, the name of the "Target Variable" is requested (see Figure 3.4). This is the name of the new total score that you are going to calculate. For the sake of our example, we will name this TWLTH. The mouse indicator goes automatically to the field under the "Target Variable" heading. Type TWLTH there.

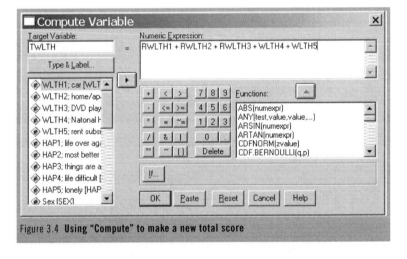

Figure 3.4 **Using "Compute" to make a new total score**

Numeric Expression
Next, you have to indicate how this total score TWLTH must be calculated. In this case, the procedure is initiated by selecting "Numeric Expression": RWLTH1 + RWLTH2 + RWLTH3 + WLTH4 + WLTH5 (make sure that you use the recoded versions of the first three variables). This operation is performed by moving these variables to the "Numeric Expression" field in the usual manner and clicking the + button to place a + sign between the variables. It is, however, easier just to type the command right in:
> Numeric Expression - type: RWLTH1 + RWLTH2 + RWLTH3 + WLTH4 + WLTH5

Once you have completed this entry, click "OK". If everything has been done properly, the new variable 'TWLTH' then appears in the data matrix.
> OK ⊙

62

You might think that TWLTH is a fanciful name: 'wlth' is, of course, an abbreviation for wealth, while 't' indicates that a total score is involved. You can use any given name. In Review 3.3, we summarize the procedures for combining data.

You also have to compute a total score for *happiness* (THAP) by summing the values of the variables HAP1, RHAP2, HAP3, RHAP4 AND RHAP5. Give it a try. If you also want to save the variables TWLTH and THAP (for example as 'a:\data3'), you must execute a save procedure by following the steps discussed in sections 2.5 and 3.2.

Review 3.3
How do I combine data to compose a total score?

> Transform ⊙
> > Compute... ⊙
> > > Target Variable - type (for example): New variable
> > > Numeric Expression - type: VAR1 + VAR2 + VAR3 + VAR4 + VAR5 ... VAR_n
> > > OK ⊙

3.5 How can I take missing values into account when combining data?

As shown in Figure 3.7, the missing elements in the scores of Respondent 2 include the variables rwlth1 and rwlth2, as the scores for wlth1 and wlth2 for this respondent are missing. When a data element is missing for one respondent, the TWLTH variable for the same respondent is marked by a space or a blank. This therefore means that the total score for Respondent 2 is omitted. For Respondent 2, TWLTH has the value ','. You can verify this by having the programme calculate the frequency distribution of the TWLTH (see Figure 3.5).

You can see in the printout in Figure 3.5 that as many as 103 respondents failed to respond to one or more of the wealth items and that all these respondents do not have any TWLTH score.

If many respondents do not answer some of the questions, there would be hardly any total scores that would remain. Consequently, much information is lost. Frequently, it only involves one or two unanswered questions, while most of the items are completed. In such circumstances, it would be better for you to calculate the *average or mean score* for the variables WLTH1 to WLTH5 instead of the total score. We will again compute examples based on the five statements concerning wealth. The average score will be named MWLTH, 'M' standing for mean. We will once more make use of the "Transform" and "Compute" procedures (see Figure 3.6).

63

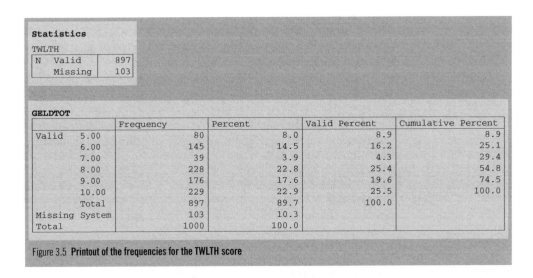

Statistics

TWLTH

N	Valid	897
	Missing	103

GELDTOT

		Frequency	Percent	Valid Percent	Cumulative Percent
Valid	5.00	80	8.0	8.9	8.9
	6.00	145	14.5	16.2	25.1
	7.00	39	3.9	4.3	29.4
	8.00	228	22.8	25.4	54.8
	9.00	176	17.6	19.6	74.5
	10.00	229	22.9	25.5	100.0
	Total	897	89.7	100.0	
Missing	System	103	10.3		
Total		1000	100.0		

Figure 3.5 **Printout of the frequencies for the TWLTH score**

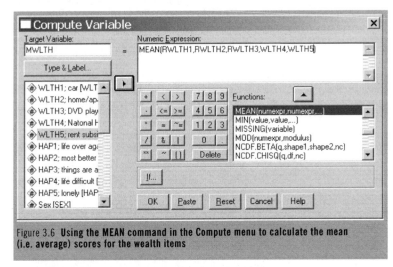

Figure 3.6 **Using the MEAN command in the Compute menu to calculate the mean (i.e. average) scores for the wealth items**

MEAN

Type the name of the new variable in the "Target Variable" field: MWLTH. Figure 3.6 shows that, in addition to summation, multiplication and the like, it is possible to execute many other arithmetic operations, all of which are displayed in the "Functions" box. The list begins with "ABS", which you would use to compute absolute values. Using the arrow button to scroll down, you will find the entry for the "MEAN" function. By double clicking it, the MEAN command appears in the "Numeric Expression" window. You only have to indicate the variables for which you need to calculate the mean score by inserting them in the parentheses. Hence, in our case: rwlth1, rwlth2, rwlth3, wlth4 and wlth5. By clicking "OK", you create a new variable, MWLTH, for which the value is a respondent's mean score for wealth. Hence:

64

> Transform ⊙
> Compute... ⊙
> Target Variable - type (for example): MWLTH.
> Functions window
> MEAN ⊙
> ▶ [to the Numeric Expression field] ⊙

Next, type the names of the variables for which the mean must be calculated, for example: rwlth1, rwlth2, rwlth3, wlth4 and wlth5.

> OK ⊙

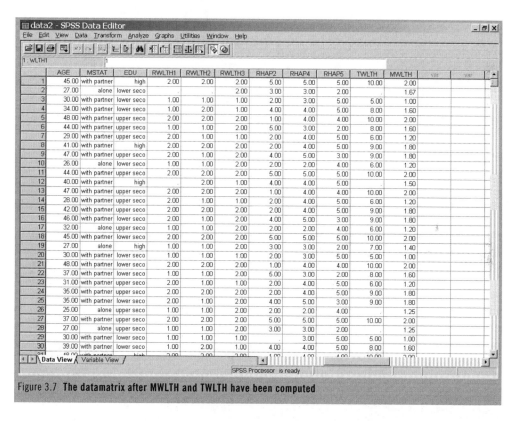

Figure 3.7 **The datamatrix after MWLTH and TWLTH have been computed**

In Figure 3.7, Respondent 2 has a mean score listed under *mwlth* of 1.67 but has no total sum of scores indicated in the *twlth* column. The three responses supplied by Respondent 2 had the scores 2 (rwlth3), 1 (wlth4) and 1 (wlth5). Their mean is 1.67, as is also shown in the data matrix. If you wish to produce a frequency printout of MWLTH, you will discover that there are no longer any missing values.

It is again prudent to save your new data matrix under a new name, for example 'a:\data4' and to note this update in your logbook. Recode then only the happiness variables 2, 4, and 5. Change a value 5 into a 1, a value 4 into a 2, etc. A high score for 65 all items then indicates a relatively high level of happiness. Next, add the happiness variables together to arrive at a total score identified, for example, as 'THAP'.

imputation of the series mean

As a 'solution' to missing data, it is also possible to employ the procedure known as 'imputation of the series mean'. The respondents who have an omitted score for a variable are given the mean score of the remaining respondents for that same variable. (There are also other techniques to attribute values to missing data, but they fall outside the scope of this discussion.) SPSS can be used to calculate the arithmetic series mean. Although there are several fundamental objections associated with this procedure, we will not discuss them any further here.
The operation can nevertheless be executed as follows:
> Transform ⊙
> Replace Missing Values... ⊙

Replace Missing Values

Darken the variables of which the missing values are to be replaced by the series mean, for example:
> RWLTH1, RWLTH2, RWLTH3, WLTH5, WLTH5
> ▶ [move to the Numeric Variable → Output Variable field] ⊙

You conclude this procedure by clicking:
> OK ⊙

As shown in Figure 3.8, SPSS automatically makes a new variable from the imputed series mean; for example, RWLTH1 becomes RWLTH1_1. In the data matrix (Figure 3.9), you will note that the scores for the new TWLTH variable do not deviate very much from the product of MWLTH multiplied by five (there are, after all, five wealth items). These new variables can be saved to a diskette in a file named 'a:data4.sav'.

You can only use either technique to solve the problem of missing values if there are not too many omissions by too many respondents.

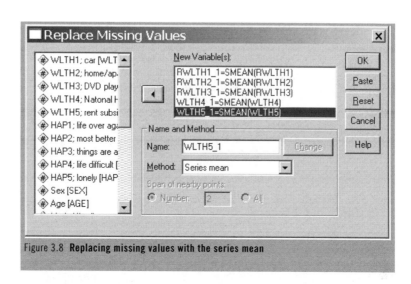

Figure 3.8 **Replacing missing values with the series mean**

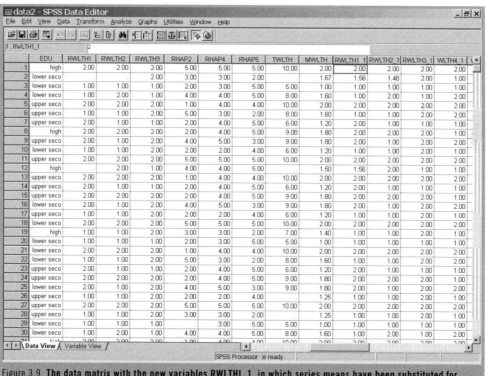

Figure 3.9 **The data matrix with the new variables RWLTHI_1, in which series means have been substituted for missing values**

3.5 How can I take missing values into account when combining data?

The examples of the *compute procedures* make it clear that you can do more with them than just calculate a sum. They can also be used to execute all types of arithmetic functions. Furthermore, you can enter commands for various operations to be performed in combination with each other.

The most important arithmetic functions that can be performed by SPSS are listed below, along with the symbols used to designate them:

sum (addition)	+
subtraction	−
product (multiplication)	*
division	/

Key words

3.1
- recoding (Transform > Recode)
- recode into same variables
- recode into different variables
- Old and New Values
- Add

3.2
- saving recoded values (Save as)

3.3
- classes / categories
- range: lowest through ...
- range: ... through highest

3.4
- new variable (Transform > Compute)
- target variable
- numeric expression

3.5
- mean
- imputation of series mean (Transform > Replace Missing Values)
- sum / addition
- subtraction
- product / multiplication
- division

H3 How do I modify or combine data?

4

How do I verify the homogeneity of the composite scores?

Money couldn't buy friends
but you got a better class of enemies.

Spike Milligan

1	2	3	4	5	6	7
How do I prepare myself to work with this Introduction?	How do I enter my data in the computer?	How do I modify or combine data?	How do I verify the homogeneity of the composite scores?	How do I analyse my data when a frequency research question is involved?	How do I analyse my data when a comparative research question is involved?	How do I analyse my data when a correlation research question is involved?

4.1
How can I examine if a composite score or scale is homogeneous?

4.2
How can I determine which items are good and which are bad?

A number of complex concepts, such as ambition, neuroticism
and, of course, happiness, are frequently measured by employing
various indicators for the concept and combining them to
establish a total score. In this book, we have, for example,
measured the notion 'wealth' by formulating five statements
concerning specific characteristics that translate the concept of
'wealth', such as the statement: I own a car. The respondent can
then indicate if this is or is not the case. We would like to
combine the responses to the five wealth items into a total score
for wealth by summing the scores of each individual item. To be
justified in doing this, you first have to check if all the items
measure the same thing: that is, if they are homogeneous. In
Section 4.1, we explain how you run an item analysis in SPSS to
verify the homogeneity of scores. The result of the item analysis,
however, is not only a figure indicating the degree of
homogeneity (Cronbach's alpha). You can also determine which
items are and are not adequate by computing the item-total
correlation. On the basis of this calculation, you can eliminate
bad items, so you are left with a homogeneous test (Section 4.2).

Prior knowledge
◼ We assume that:
- you can start SPSS for Windows (1.4);
- you can retrieve a data file (2.6);
- you can recode a variable's values (3.1).

4.1 How can I examine if a composite score or scale is homogeneous?

When more than one indicator has been used to measure a concept like happiness, you will usually want to combine them into a total score. We have operationalized (translated) happiness, or rather the satisfaction with this life that an individual leads, into the five statements or items listed in the introduction to Chapter 1. For the sake of completeness, we have mentioned them again here. The five statements are:

- If I could live my life over again, I would (certainly not – definitely) live it in the same way [HAP1].
- Most other people are (certainly not – definitely) better off than I am [HAP2].
- Things (are certainly not – exactly) as I would like them to be [HAP3].
- Life is (certainly not – extremely difficult) [HAP4].
- I (certainly do not – definitely) feel lonely [HAP5].

total score
index

In most such cases, the responses to the individual items are summed to calculate a total score or index. In this case, the total score for 'happiness' varies from a score of 5 (which is 5 times code 1) to a maximum of 25. You can check to see if it is permissible simply to sum the individual scores into a total score by determining the reliability or homogeneity of this scale (that is, the extent to which the selected items measure the same characteristic). Based on the contents of the five statements, you can assume that they measure the same construct; that is, people's happiness about their lives. However, you will have to verify this assumption. The extent to which a number of items measure the same construct is mostly expressed as a number, the homogeneity coefficient alpha (also known as the alpha reliability index). The alpha can have a value ranging from 0.00 to 1.00. A value of 0.00 indicates a total lack of correlation among the items; 1.00 means that the items overlap each other completely.

In SPSS, you use the "Reliability" procedure to check if the proposed 'happiness' scale is reliable or homogeneous. If the scale is not homogeneous, you can use the programme to determine which statement or statements do not measure the same construct as the others and are therefore causing a low level of homogeneity.

Note: if you wish to determine the homogeneity of a scale, all the variables must be scaled in the same direction. In our example, a high score on all happiness items must correspond to a high(er) degree of satisfaction with life. The same holds true for the items concerning wealth. For the procedure of recoding variables, see Section 3.1. The file containing the recorded happiness variables was stored in 'a:\data4'; the recoded variables were labelled RHAP2, RHAP4 and RHAP5. In this file, we also recoded the first three wealth items (WLTH1, WLTH2 and WLTH3) so that '1' was an indication of less and '2' of more wealth. The new recoded wealth variables are: RWLTH1, RWLTH2 and RWLTH3.

To locate the "Reliability" command, we select "Analyze" in the main menu and then "Scale" from the roll down menu that appears. In the menu that now appears on the screen, we find a list of commands including Reliability. Select the latter and click it once:

> Analyze ⊙
  > Scale
    > Reliability Analysis ⊙

See Figure 4.1.

You must next indicate the variables for which the homogeneity has to be calculated. Select the variables HAP1, RHAP2, HAP3, RHAP4 and RHAP5 and place them, following the usual procedure, in the field beneath the "Items" button.

> HAP1 ⊙
> ► [to the Items field] ⊙
> etc. through RHAP5

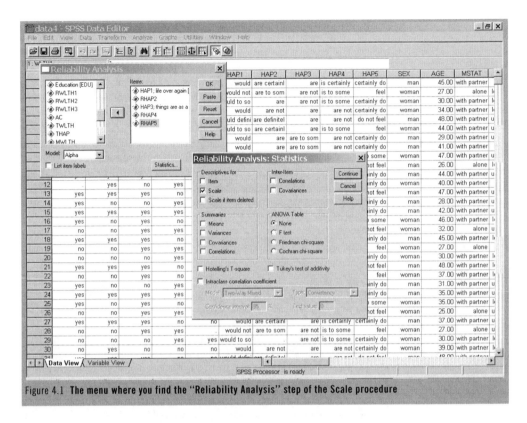

Figure 4.1 **The menu where you find the "Reliability Analysis" step of the Scale procedure**

Cronbach's alpha

Model

Split-half method

By default, SPSS calculates the reliability measure, Cronbach's alpha. This is the most widely used reliability statistic in classical test theory. You also see this appear in the left window beside "Model" (normally, the word "Alpha" appears here). If you wish, you can also use other standards of measure, for example the Split-half method. You then need to continue clicking the down arrow in the "Model" field until "Split-half" appears. We would like to use Cronbach's alpha to examine the homogeneity of the happiness items and, therefore, do not need to change anything, but we still have to mark the "Scale item" in the "Descriptives for" box on the "Statistics" submenu. The procedure is executed in the following way:

> Statistics ⊙
> place a check beside Scale ⊙
> Continue ⊙
> OK ⊙

The output reveals that the homogeneity coefficient alpha is calculated using the scores for the responses to the five statements provided by 951 respondents. Forty-nine respondents skipped over one or more items and are removed from the analysis by the programme. You can see that the homogeneity of the items is 0.58, which is not so high. The maximum is 1.00 (see Figure 4.2).

72

H4 How do I verify the homogeneity of the composite scores?

The five items do indeed measure something in common, but there is quite a bit of discrepancy among them. In Section 4.2, we explain how you can determine if there are items that reduce the homogeneity, as it would be better for us not to include these in the analysis.

```
RELIABILITY ANALYSIS - SCALE (ALPHA)

Statistics for      Mean    Variance    Std Dev    N of Variables
SCALE            17.9054    10.5131      3.2424                 5

Reliability Coefficients

N of Cases = 951.0              N of Items = 5

Alpha = .5799
```

Figure 4.2 **Output from the "Reliability Analysis" step of the "Scale" procedure**

Review 4.1
How can I check if my scale is reliable?

> Analyze ⊙
> > Scale
> > > Reliability Analysis ⊙
> > > > VAR1 ⊙
> > > > ▶ [to the Items field] ⊙
> > > > etc. through VARn
> > > > Statistics ⊙
> > > > > place a check beside Scale ⊙
> > > > > Continue ⊙
> > > > OK ⊙

4.2   How can I determine which items are good and which ones are bad?

In our example, the alpha is 0.58, which is rather low. When a scale is completely homogeneous, the coefficient is 1.00. For a totally heterogeneous scale, it is 0. An alpha of 0.58 means the happiness total score (THAP) is only passably homogeneous. If the scale adequately measures happiness (if it has validity) is another question. Homogeneity is a characteristic of the reliability of a scale and does not say anything about the validity. However, a (high) degree of reliability is a condition for validity. Depending on the complexity of the construct to be measured, it is usually assumed that a minimum alpha of 0.60 is desirable for complex constructs, such as neuroticism and intelligence, and of 0.80 for less complex constructs, such as numerical ability. The 0.58 alpha that we have computed can be identified as mediocre (neither high nor really low).

validity

73

The SPSS programme offers a number of possible ways to obtain more specific statistical information about scale quality. With an alpha of 0.58, you are of course curious to know why this calculated alpha is so relatively low. If you have just worked through Section 4.1, the procedure will still be familiar to you. For the sake of completeness, however, we will repeat it here.

> Analyze ⓥ
> > Scale
> > > Reliability Analysis ⓥ
> > > HAP1 ⓥ
> > > ▶ [to the Items field] ⓥ
> > > etc. through RHAP5

Statistics
Scale if item deleted

Now select Statistics and then the "Scale if item deleted" procedure (top left corner in the "Reliability Analysis: Statistics" window). Once you have made this selection, a check appears in the box beside the designated procedure:

> Statistics... ⓥ
> > Scale if item deleted ⓥ

The procedure is executed in the following way:

> Continue ⓥ
> OK ⓥ

The results of the reliability analysis are found in Figure 4.3.

```
RELIABILITY ANALYSIS - SCALE (ALPHA)
```

Item-total Statistics

	Scale Mean if Item Deleted	Scale Variance if Item Deleted	Corrected Item-Total Correlation	Alpha if Item Deleted
HAP1	14.2755	6.6082	0.4958	0.4301
RHAP2	14.9516	8.1998	0.0429	0.7369
HAP3	14.4721	7.4326	0.6165	0.4230
RHAP4	14.0757	6.4932	0.7791	0.3196
RHAP5	13.8465	8.4417	0.1270	0.6401

Reliability Coefficients

N of Cases = 951.0   N of Items = 5

Alpha = 0.5799

Figure 4.3 **Output from the "Reliability" procedure requesting the Statistic "Scale if item deleted"**

With the "Scale if item deleted", you obtain statistical information about the correlation between a given individual variable (or statement) and the total score of the remaining items. It is useful to determine the extent to which one item measures the same thing as the remaining ones do. The printout (Figure 4.3) shows that recoded Item 4 ("Life is difficult") measures, to a great extent, the same thing as the remaining items. After all, the correlation with the remaining items is 0.78. The second happiness item

74

("Most other people are better off than I am") displays, conversely, hardly any overlap with the remaining items; the correlation is 0.04. It is therefore sensible to eliminate this item from the scale. It clearly measures something else. In the printout, you can also see what effect the elimination of an item has on reliability. The 'Alpha if Item Deleted' column indicates that the homogeneity would become 0.74 if RHAP2 were to be removed from the list. The homogeneity would rise considerably as a result. If you also leave RHAP5 out, the homogeneity increases somewhat more. Although you then only have a few remaining items, it is nevertheless best to scrap items RHAP2 and RHAP5 and to continue with items HAP1, HAP3 and RHAP4. With these three items, you retain a homogeneous whole. It is of course advisable to scrutinize the content of the items once more. A close examination of RHAP2 and RHAP5 reveals that their content deviates somewhat from the other items.

Alpha if Item Deleted

We now have to compute the homogeneity for the three-item scale again. Try it yourself and compare your results with those in Figure 4.4.

```
RELIABILITY ANALYSIS - SCALE (ALPHA)
```

Item-total Statistics				
	Scale Mean if Item Deleted	Scale Variance if Item Deleted	Corrected Item-Total Correlation	Alpha if Item Deleted
HAP1	7.2439	1.9730	0.7661	0.7204
HAP3	7.4431	3.1829	0.6211	0.8437
RHAP4	7.0610	2.6840	0.7537	0.7184

```
Reliability Coefficients

N of Cases = 984.0          N of Items = 3

Alpha = 0.8337
```

Figure 4.4 **Output from the "Reliability" procedure for items HAP1, HAP3 en RHAP4**

On the basis of previously mentioned criteria, an alpha reliability index of 0.83 is said to be rather high, especially when such a small number of items is involved. In calculating the homogeneity, you must certainly be aware of the fact that a scale's homogeneity is higher when it contains more items. If you have a large number of items the homogeneity can still be quite high despite relatively low item-total correlations.

If you take another look at Figure 4.1, you will see that the "Statistics" submenu allows you to compute more than just the item-rest correlation and the 'Alpha if Item Deleted'. You can, for example, have SPSS calculate the mean and the standard deviation of the individual items, as well as of the total score. Here we have only discussed the most important statistics.

How do I obtain statistical information about the quality of a composite score or scale?

> Analyze ⊚
  > Scale
    > Reliability Analysis ⊚
      > VAR1 ⊚
      > ▶ [to the Items field] ⊚
      > etc. through VARn
      > Statistics... ⊚
        > Scale if item deleted ⊚
        > Continue ⊚
        > OK ⊚

The computer only performs a reliability analysis and does not automatically calculate each respondent's total score on the happiness scale. You will have to decide the implications associated with the results of this analysis for yourself. For example, you could, nevertheless, decide to keep certain items for reasons of content and be satisfied with a slightly lower degree of reliability. If you would like to use the total score for happiness in subsequent analysis, you will first have to use the "Compute" procedure to make a new variable (Section 3.4). We have chosen THAP as the name for the target variable comprising the sum of the scores for the three remaining items (HAP1, HAP3 and RHAP4). It is sensible to save this variable. In our case, the file now becomes 'a:\data5'. Any subsequent work, such as that involving the calculation of the relationship between wealth and happiness, will not use the individual items but the score on the Happiness scale: THAP.

You can employ the same procedure to arrive at a total score for the wealth variables. We have also done this and, based on an item analysis, decided on a scale comprising items 2, 4 and 5 (more accurately: RWLTH2, WLTH4 and WLTH5; see Section 3.5). The Cronbach's alpha for the scale consisting of these three items is 0.81. The results from this analysis can be found on the website in the file entitled 'OUTPUT4'. Try to calculate it for yourself as well. Again, we used the compute procedure to sum the scores for the three wealth items. The total score is named 'TWLTH' and written over the old set of 'TWLTH' variables, which had been computed on the basis of all five wealth items. Thus a revised set of variables can also be found in 'a:\data5'.

Key words

4.1
■ composite score
■ total score/index
■ reliability, homogeneity
■ homogeneity coefficient alpha
■ reliability (Analyze > Scale > Reliability)

■ Cronbach's alpha
■ split-half method
4.2
■ statistics
■ scale if item deleted
■

76

5

How do I analyze my data when a frequency research question is involved?

> *Money, it turned out, was exactly like sex.*
> *You thought of nothing else if you didn't have it*
> *and thought of other things if you did.*

James Baldwin

1	2	3	4	5	6	7
How do I prepare myself to work with this Introduction?	How do I enter my data in the computer?	How do I modify or combine data?	How do I verify the homogeneity of the composite scores?	How do I analyse my data when a frequency research question is involved?	How do I analyse my data when a comparative research question is involved?	How do I analyse my data when a correlation research question is involved?

5.1	5.2	5.3	5.4
The making. reading, presenting, interpreting and reporting of a frequency distribution	How do I calculate the frequencies of subgroups?	How do I compare (sub)groups?	How do I make variables comparable?

If you need to know how often something occurs, you can find this out by composing a frequency distribution. The appearance of such a distribution will depend on the level of measurement involved with your variables. Construction of a frequency distribution sometimes makes little sense, especially for such variables as age, which have a lot of values. You can better organize the data by constructing a table that displays both frequencies and percentages, or with a stem-and-leaf diagram. The latter is a combined table and graph that enables you to quickly see which values occur most and least frequently. You can also use SPSS to calculate all kinds of data-set characteristics, such as the mean, the median (middle value), the mode (the most frequently occurring value), the standard deviation and the variance. If, for example, you compute the mean for a sample, it is mostly used to estimate the population mean. Given that various sample means will, in each instance, diverge somewhat from each other, you consequently commit a so-called estimation error. Therefore, you have to indicate the margins within which the population mean is very likely to fall (standard error of mean). We explain how you use SPSS to do all these operations in Section 5.1.1. SPSS also makes it possible to represent frequencies in a diagram. When you are using nominal variables with only a few values, you can employ, for example, a pie or bar chart. When your research involves a continuous variable containing relatively many values, it is better to construct a histogram. This makes it possible for you to determine if the data has a normal distribution (Section 5.1.2). In sections 5.1.3 and 5.1.4, we discuss how you should read and interpret the frequency output from SPSS and, in Section 5.1.5, how you should report it. Sometimes you do not just want to know the frequency for the complete group but also for various subgroups, such as men or women. SPSS also enables you to calculate the data for such subgroups (Section 5.2). There are also graphic techniques, like the clustered bar chart and the box

Prior knowledge

■ We assume that:

• you can start SPSS for Windows (1.4);

• you can retrieve a data file (2.6);

• you can distinguish among the various levels of measurement (1.2.2);

• you know what discrete and continuous variables are (1.2.2);

• you know what a normal distribution is (1.3);

78

plot, that make it possible to compare groups (5.3). Comparisons on the level of the individual can be undertaken by converting scores having different means and standard deviations into standard scores. We describe how SPSS can be used to do this in Section 5.4.

5.1 The making, reading, presenting, interpreting and reporting of a frequency distribution

5.1.1 What is a frequency distribution?

frequency
distribution

A frequency distribution summarizes a variable's scores. It shows how frequently each value or category of a variable occurs, expressing their frequency as absolute numbers and/or percentages. In our example, you can state that there were 500 women and 500 men, and, in percentages, 50% men and 50% women. For variables, like age, involving a large number of values, things are a little more complicated. The result is then a long enumeration: 31 twenty- five-year olds, 29 twenty-six-year olds, etc. Another method, which would also give your observations a bit more structure, involves the use of a stem- and-leaf diagram, an example of which is found in Figure 5.1.

stem- and-leaf
diagram

```
Frequency    Stem  &   Leaf
    31.00      2  .   555555555555555
    57.00      2  .   6666666666666677777777777777
    64.00      2  .   8888888888888888999999999999999999
    77.00      3  .   00000000000000000000001111111111111111111
    58.00      3  .   22222222222222223333333333333
    72.00      3  .   444444444444444444444555555555555555
    48.00      3  .   666666666777777777777777
    68.00      3  .   8888888888888888999999999999999999999
    75.00      4  .   0000000000000000000000011111111111111111
    68.00      4  .   2222222222222222222233333333333333333
    63.00      4  .   44444444444444445555555555555555
    66.00      4  .   6666666666666677777777777777777777
    52.00      4  .   8888888888888889999999999
    48.00      5  .   000000000000111111111111
    71.00      5  .   222222222222222222223333333333333333
    70.00      5  .   4444444444444444455555555555555555555
```

Stem width: 10.00
Each leaf: 2 case(s)

Figure 5.1 **Age (in years): Stem-and-Leaf Plot**

The stem lists age in decades. The leaves indicate the individual years of age between each decade. In the first row, you can see therefore that there are 31 people aged 25. The second row lists 57 twenty-six and twenty-seven-year olds. A '7' appears 14 times, which means that there are 14 * 2 = 28 people who are 27 years   79 old (as, in this case, each leaf counts for two observations). The advantage of the stem-and-leaf diagram is that you immediately obtain an idea of the age distribution, as you see at a glance that there is a roughly equal representation of people from all age groups.

It is more usual to structure the data a little by placing it in a table. And if there are many values, it is useful to divide the table **classes** into classes. The number of classes depends on the number of observations. If not so many people have participated in your study, it would be better if you did not make the classes too small. The example involves 1000 people, so you can, in this case, certainly decide to make relatively small age classes encompassing five years, for example (see Table 5.1).

Table 5.1 **Age distribution of the sample group**

Age in years	Frequency	Percentage	Cumulative percentage
25 - 29	152	15.4	15.4
30 - 34	176	17.8	33.2
35 - 39	147	14.9	48.1
40 - 44	173	17.5	65.6
45 - 49	151	15.3	80.9
50 - 54	151	15.3	96.2
≥ 55	38	3.8	100
Total	988		

As shown in the table, not all the ages of the 1000 people involved in the study are known. In the example, the percentage (in this **Valid Percent** case labelled in SPSS as the "Valid Percent") is calculated for the 988 individuals whose ages are known. The final column also pro- **cumulative** vides a record of cumulative percentage: 48.1% of the partici- **percentage** pants are younger than forty years old.

If there are many different values, short, encapsulating features characterising a series of figures or scores are often employed. **mean** Most commonly, the mean is used, which, in our case, is the aver- age age. The mean of a series of numerical data is the sum of the figures divided by their number. Consequently, if the exam grades for nine students are 2, 5, 5, 6, 6, 6, 7, 7 and 8, the mean exam grade is equal to the sum of all the grades divided by the number of the students:
$(2 + 5 + 5 + 6 + 6 + 6 + 7 + 7 + 8) / 9 = 5.78$

As illustrated by the above example, the mean can be a number that does not actually occur in the series of figures or, even, in re- 80 ality.

Sometimes, other measurements are used, especially when there are some extreme scores. Such is also the case in our example. The student who received the grade of 2 deviates, to a certain extent, from the rest and causes the mean to be somewhat lower than it should be. Based on the mean grade of 5.78, you could have the impression that half the students failed the exam, which was actually not the case. When extreme scores occur, preference is often given to the median or the mode. The median is the middlemost value, when scores are ranked from lowest to highest. In our case, this is the fifth score, which is 6. The median, in our case, is consequently 6, and this figure characterizes the distribution of grades somewhat more accurately. When you have an even number of scores, for example ten, you then take the mean of the two middlemost values. Hence, with ten students, this would be the mean of the grades assigned to students 5 and 6. The mode is the most frequently-occurring score in the series, which, in our case, is also 6, for there are three sixes in contrast to two fives and two sevens.

median
mode

mode

Sometimes the mean is also misleading for other reasons and provides inadequate information. Take a look at the following example. In both cases, the mean is 7, but the distributions do not, in any way, resemble each other.

Grades for group 1   Grades for group 2
5, 6, 7, 8, 9   7, 7, 7, 7, 7
Mean = 7   Mean = 7

standard deviation
variance

Although the means are the same, the groups are in fact completely dissimilar, since the grades within the group are distributed in completely different manners. The first group displays a far greater variation of grades than the second. In statistics, variation is called the standard deviation or variance. You obtain the standard deviation by taking the root of the variance. The variance is calculated by first computing the distances that the various scores are from the mean. These distances are squared and the squares subsequently summed. For group 1, the result of this calculation is 10 and for group 2, of course, 0. These totals are divided by the total number of scores. In our example, the variance for group 1 is consequently $10/5 = 2$, and for group 2 is $0/5 = 0$. (See Table 5.2.)

The standard deviation for group 1 in our example is $\sqrt{2} = 1.41$ and for group 2 is $* \sqrt{0} = 0$.

If now, in addition to the identical means for both groups, you also indicate the standard deviations, it immediately becomes clear that the distributions within the two groups are strongly divergent.

It is most useful to work out the variance by hand, as shown in Table 5.2

81

Table 5.2 **Summary of exam grades**

	Score	Mean	Deviation	Squared deviation
	5	7	−2	4
	6	7	−1	1
	7	7	0	0
	8	7	1	1
	9	7	2	4
Totaal	35		0	10

Variance = 10/5 = 2   Standard deviation = $\sqrt{2}$ = 1.41

If you have not made any errors in arithmetic, the total for the deviation is always 0.

5.1.1.1   When do I use a frequency distribution?

A frequency distribution is used when a specific research question asks about 'how often' or 'to what extent' something occurs. Hence, for example: how many women work for the government? In constructing frequency distributions for nominal and ordinal variables (sex, education), the appropriate procedure involves a simple counting of values, although percentages may be used to provide a clearer view of the whole. The character of a frequency distribution of variables measured on the interval or ratio level, such as age and salary, is mostly reported in terms of the mean and the variance or standard deviation. If extreme scores are an issue, such as extremely high salaries for directors, the median and mode are also indicated because the extreme scores can distort the mean.

Be very careful to note if your research involves a population or a sample. If you have drawn a sample, you can also generate totals, percentages, means and the like, but you cannot simply generalize the results to the population. When using a sample, some consideration must always be given to the margin of error resulting from chance: the standard error of estimate. 40% of the men in the sample do not automatically represent 40% of the men in the population. This could also be 38% or 42%. A mean age of 31 years old in the sample does not automatically mean that the mean age in the population is also 31. When you have constructed a representative sample, the sample mean will more closely approximate the actual mean of the population if the sample is larger. The margin of error is additionally expressed in the form of the standard error (standard error of the mean). (Also see 1.3.)

standard error of estimate

standard error

82

5.1.1.2   How do I generate a frequency distribution?

In Section 2.5, we have shown how you use the "Frequencies" procedure to check if you have made any errors when entering data. In some cases, it is sufficient to compute the frequency distribution of all the variables involved in the study. In other cases, you need some more information about the variables. It is, however, possible to do much more with the "Frequencies" procedure than we have indicated in Section 2.5. Among other things, you can use it to calculate the mean and the standard deviation, also known as the variation, of each variable. You can also embellish the output by having the program make bar charts, for example, or other types of diagrams. We will discuss these elements more fully in Section 5.2.

The data that we are now going to use are stored in the data file entitled 'data5'. This file must be retrieved in the usual manner (Section 2.6).

To use the "Frequencies" procedure to construct a frequency distribution for a set of variables, you must first go to "Analyze" in the menu bar. In this roll- down menu, you select "Descriptive Statistics" and then "Frequencies":

> Analyze ⊙
>> > Descriptive Statistics
>>> > Frequencies ⊙

You must next indicate the variables for which the frequencies have to be calculated. This is done by using the mouse indicator to mark the relevant variable(s) and moving them in the usual way to the field entitled "Variable(s)". As an example, let us calculate the frequency distribution for the THAP variable, the sum of the happiness variables HAP1, HAP3 and RHAP4:

> THAP ⊙
> ▶ [to the Variable(s) field] ⊙

Using the "Statistics" button, you can indicate if you would like to have statistics computed and, if so, which ones. We choose the mean, median and mode as measures for the central tendency, as well as the standard deviation and variance as measurements of the dispersion.

> Statistics... ▼
  > Mean ▼
  > Median ▼
  > Mode ▼
  > Std. deviation ▼
  > Variance ▼
  > Continue ▼
> OK ▼

Figure 5.2 shows that still other statistics can be chosen in the "Statistics" window, such as the minimum and the range.

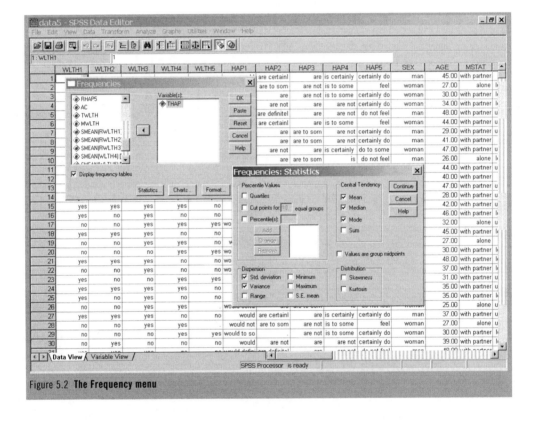

Figure 5.2 **The Frequency menu**

The results of the procedure are displayed in Figure 5.3.

Statistics

THAP

N	Valid	984
	Missing	16
Mean		10.8740
Median		12.0000
Mode		13.00
Std. Deviation		2.32995
Variance		5.42866

THAP

		Frequency	Percent	Valid Percent	Cumulative Percent
Valid	6.00	67	6.7	6.8	6.8
	7.00	79	7.9	8.0	14.8
	8.00	80	8.0	8.1	23.0
	9.00	3	0.3	0.3	23.3
	10.00	88	8.8	8.9	32.2
	11.00	172	17.2	17.5	49.7
	12.00	129	12.9	13.1	62.8
	13.00	366	36.6	37.2	100.00
	Total	984	98.4	100.00	
Missing	System	16	1.6		
Total		1000	100.0		

Figure 5.3 **Output belonging to the "Descriptives Statistics" > "Frequencies" procedure for THAP**

5.1.2 How do I make graphic representations of frequencies?

5.1.2.1 What is a pie chart, a bar chart or a histogram?

pie chart
circle diagram

A *pie chart* (also known as a circle diagram) is a circle that is divided into wedges (pie slices) with which you can graphically represent a frequency distribution. Each category or value of the variable is represented as a piece of the pie. The size of these slices reflects the relative frequency.

bar chart

A *bar chart* (or bar graph) represents a frequency distribution in the form of a series of bars or columns. The categories or values are placed along the horizontal axis (for example: single, with partner, with partner and children). The frequencies are listed on the vertical axis, hence the number of respondents who are single, etc. The bars do not contact each other (they represent discrete values) and all have the same width.

histogram

A *histogram* is comparable to a bar chart but its bars do contact each other. It is a diagram representing a frequency distribution of classes.

5.1.2.2 When do I use a pie chart, a bar chart or a histogram?

A *pie chart* is only used to represent nominal variables with a limited number of values, such as sex or marital status.

A *bar chart* can be employed to represent nominal variables and discrete numerical variables; for example, educational level with the values of lower secondary, upper secondary and higher education.

A *histogram* is meant to provide a representation of continuous variables. These are numerical variables in which all possible values may occur. Age is an example of this type of variable, given that it can take months and even weeks or days into account.

5.1.2.3 How do I create a pie chart, a bar chart or a histogram?

Output from the Frequencies procedure can be obtained in tabular form, but you can also have a frequency diagram made. You need then to choose one of the menu options displayed when the **Charts** "Charts" button is clicked (see Figure 5.4).

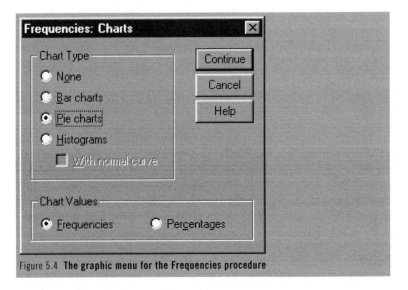

Figure 5.4 **The graphic menu for the Frequencies procedure**

circle diagram You can use the "Pie Charts" procedure to make a circle diagram from the distribution of a nominal variable. To provide an example, we have done this for the *sex* variable in the happiness study. The result is displayed in Figure 5.5, where you can see in a glance that the number of men is equal to the number of women.

To have SPSS make a pie chart from the frequency distribution for *sex* separate from the associated frequency distribution, you can also use the following procedure:

86

> Graphs ⊙
  > Pie... ⊙
    > Define ⊙
      > sex ⊙
      > ▶ [to the Define Slices by field] ⊙
      > OK ⊙

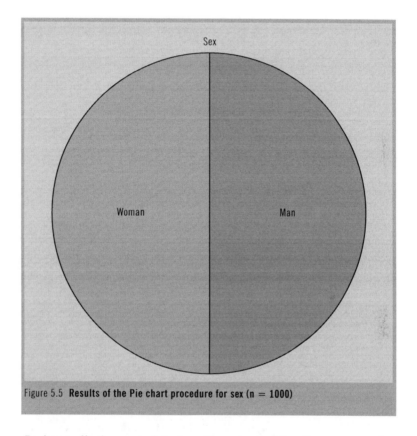

Figure 5.5 **Results of the Pie chart procedure for sex (n = 1000)**

In Appendix 2, we explain how the output of the diagrams can be modified in order to change colours, to add text and so on.

bar graphs

The "Bar charts" procedure can be used to represent the frequency distribution of nominal and ordinal variables in bar graphs. It is helpful and illustrative to execute such supplementary procedures in association with the "Frequencies" procedure when there are not so many values or categories, as is the case for sex, marital status and the like.

Figure 5.6 shows the results of such a procedure involving the distribution of the *marital status* variable. In a glance, you can also see here that single persons constitute a minority. To have SPSS make a bar chart from the frequency distribution for the *marital status* variable separate from any procedure to calculate the frequency distribution, you can also use the following procedure: **87**

> Graphs ▼
    > Bar... ▼
        > Define ▼
            > marital status ▼
            > ▶ [to the Category Axis field] ▼
            > OK ▼

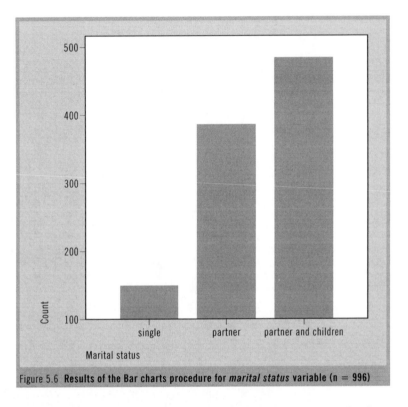

Figure 5.6 **Results of the Bar charts procedure for *marital status* variable (n = 996)**

histogram The "Histograms" procedure can be used to represent the frequency distribution of a continuous variable in histogram form. You can also add something extra to this diagram by placing a check in the box labelled "With normal curve"; see Figure 5.4. You are then not only provided with the histogram, but also the distribution that the histogram must have in order to be a normal distribution. For instructional purposes, we used the discrete variable THAP to create an example. The results are displayed in Figure 5.7. To have SPSS make a histogram from the frequency distribution for the THAP variable separate from the associated frequency distribution, you can also execute the following procedure:

> Graphs ▼
   > Statistics... ▼
   > THAP▼
   > ▶ [to the Variable field] ▼
   > Display normal curve ▼
   > OK ▼

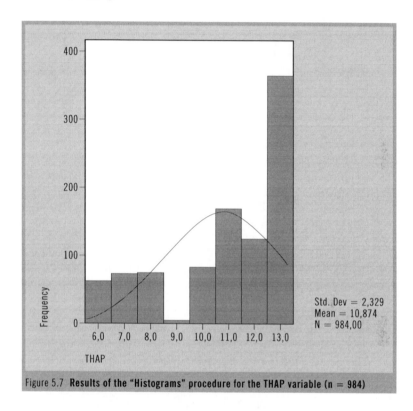

Figure 5.7 **Results of the "Histograms" procedure for the THAP variable (n = 984)**

5.1.3 How do I read the output of the Frequencies procedure?

The frequency table in Figure 5.3 consists of five columns: Valid, Frequencies, Percent, Valid Percent, and Cumulative Percent.

Valid
The 'Valid' column contains the values belonging to the THAP variable; they vary from 6 through 13. As you probably already know, they represent the sum of three happiness variables that all have values between 1 and 5. Therefore, the total score can, in principle, vary from 3 to 15. Admittedly, the extreme values of 3, 4, 5, 14 and 15 do not, in practice, occur.

Frequencies
The 'Frequencies' column indicates the number of times that a given value occurs; therefore, 67 people have a THAP score of 6; 79 people a value of 7, etc.

89

| Percent | Under 'Percent' is the percentage of people having the given THAP score out of all the people who participated in the study (= 1000). 6.7% (67 of 1000 people) have a score of 6 and 7.9% have a score of 7, etc. |

Percent

Under 'Percent' is the percentage of people having the given THAP score out of all the people who participated in the study (= 1000). 6.7% (67 of 1000 people) have a score of 6 and 7.9% have a score of 7, etc.

Valid Percent

'Valid Percent' is conversely the percentage of the given THAP score from all the people who actually responded to the relevant item (984); 6.8% (67 of the 984 people) have a score of 6 and 8.0%

Missing System

a score of 7. The individuals who did not respond or did not validly complete the item (Missing System = 16) are therefore being disregarded here.

Cumulative Percent

As previously stated, the 'Cumulative Percent' is computed by adding all the valid percentages together: 6.8% have a score of 6; 14.8% (= 6.8 + 8.0) have a score of 7 or lower; 23% (= 6.8 + 8.0 + 8.1) have a score of 8 or lower, etc.

It should further be noted that the values for the mean (10.87), the median (12.00) and the mode (13) are somewhat different. This is the case because the distribution of the THAP scores is

skewed distribution
negatively-skewed distribution
positively-skewed distribution

not a normal one (Figure 5.7). Such a skewed distribution, which includes relatively high scores, is called a negatively-skewed distribution. If there had, in fact, been a relatively large number of low scores, it would have been a positively-skewed distribution. You can therefore conclude that, in our case, relatively more people scored high for happiness. There are many people who claim to be (very) happy: the median is as high as 12, while the maximum score is 15. More than 50% of the people in the sample have a happiness score of 12 or higher.

5.1.4 How can I present a frequency distribution?

It would be unusual to include the entire output for an SPSS frequency distribution, as illustrated in Figure 5.3, in a research report. It is necessary to eliminate irrelevant information and to present an attractive summary in a tabular form for your report. We have converted the output in Figure 5.3 into a presentable format which we have labelled Table 5.3. As shown in the table, percentages are limited to two-digit figures. They are rounded off in the conventional manner, as a result of which possible rounding-off errors could occur.

THAP involves 8 values; they can all be placed in the table. If there are many values involved, a table becomes overly complicated, and it would be better for you to present your results in a bar graph or histogram.

90

Table 5.3 **The distribution of the total happiness score (THAP), the sum of the happiness variables HAP1, HAP3 and RHAP4 (in absolute numbers and percentages)**

THAP	N	%
6	67	7
7	79	8
8	80	8
9	3	0,3*
10	88	9
11	172	18
12	129	13
13	366	37
Total	984	100

* This percentage has not been rounded off, unlike the other percentages.

5.1.5 How should I interpret and report a frequency distribution?

In Section 5.1.2, we suggested that the THAP data were noteworthy because the mean happiness score is very high while the distribution is strongly skewed; there is a relatively large number of people with a high score for happiness. You could describe the results as follows:

"The mean total happiness is high (10.87), especially given the range from a minimum score of 5 to a maximum of 15. The majority of people declare themselves to be (very) happy; more than half have a score of 12 or higher."

Review 5.1

How do I calculate a frequency distribution?

> Analyze ⊙
    > Descriptive Statistics
        > Frequencies ⊙
            > VAR1 ⊙
            > ▶ [to the Variable(s) field] ⊙
            > VAR2 ⊙ etc. through VARn
            > Statistics... ⊙
                > Mean ⊙
                > Median ⊙
                > Mode ⊙
                > Std. deviation ⊙
                > Variance ⊙
                > Continue ⊙
            > Charts... ⊙
                > Pie charts ⊙

or
> Bar charts ⊙
or
> Histograms ⊙
> Continue ⊙
> OK ⊙

5.2   How do I calculate the frequencies of subgroups?

In considering the group as a whole, we have observed that the majority of people claim to be happy. Sometimes you want to know if such is the case for subgroups considered separately. How happy are men or women, for example? To discover this, you will need to have separate frequency distributions for men and

Select Cases   women. This can be done by using the "Select Cases" procedure found in the roll-down "Data" menu.

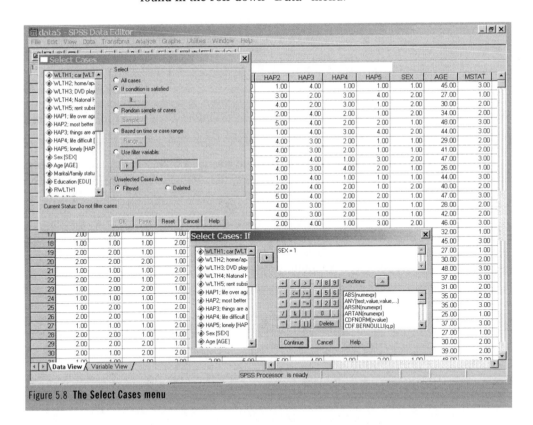

Figure 5.8 **The Select Cases menu**

As you can see, there is a number of ways to select Cases. The most common manner involves the "If condition is satisfied" procedure:
> Data ⊙
> Select Cases ⊙
> If condition is satisfied ⊙

H5 How do I analyse my data when a frequency research question is involved?

This is also the choice that has been made in Figure 5.8. You only have to indicate the condition under which you want to make the selection by clicking the "If" button. The "Select Cases: If" frame then appears. In it, you type 'sex = 1' for example, which means that the analysis is to be performed for the male group only (the code 1 is in this case standing for 'man'). The procedure is as follows:

> If… ▼
  > sex ▼
  > ▶ [to the window] ▼
  > type after 'sex': = 1
  > Continue ▼
> OK ▼

Figure 5.9 **Datamatrix for which "Select Cases: if" has been used to choose only male respondents**

	WLTH1	WLTH2	WLTH3	WLTH4	WLTH5	HAP1	HAP2	HAP3	HAP4	HAP5	SEX	AGE	MSTAT
1	1.00	1.00	1.00	2.00	2.00	4.00	1.00	4.00	1.00	1.00	1.00	45.00	3.00
2			1.00	1.00	2.00	2.00	3.00	2.00	3.00	4.00	2.00	27.00	1.00
3	2.00	2.00	2.00	1.00	1.00	3.00	4.00	2.00	3.00	1.00	2.00	30.00	2.00
4	2.00	1.00	2.00	2.00	2.00	4.00	2.00	4.00	2.00	1.00	2.00	34.00	2.00
5	1.00	1.00	1.00	2.00	2.00	5.00	5.00	4.00	2.00	2.00	1.00	48.00	3.00
6	2.00	2.00	1.00	2.00	2.00	3.00	1.00	4.00	3.00	4.00	2.00	44.00	3.00
7	1.00	2.00	2.00	1.00	1.00	4.00	4.00	3.00	2.00	1.00	1.00	29.00	2.00
8	1.00	1.00	1.00	1.00	2.00	4.00	4.00	3.00	2.00	1.00	1.00	41.00	2.00
9	1.00	2.00	1.00	2.00	2.00	4.00	2.00	4.00	1.00	3.00	2.00	47.00	3.00
10	2.00	2.00	1.00	1.00	1.00	1.00	4.00	3.00	4.00	2.00	1.00	26.00	1.00
11	1.00	1.00	1.00	2.00	2.00	4.00	1.00	4.00	1.00	1.00	1.00	44.00	3.00
12		1.00	2.00	1.00	2.00	4.00	2.00	4.00	2.00	1.00	2.00	40.00	2.00
13	1.00	1.00	1.00	2.00	2.00	5.00	5.00	4.00	2.00	2.00	1.00	47.00	3.00
14	1.00	2.00	2.00	1.00	1.00	4.00	4.00	3.00	2.00	1.00	1.00	28.00	2.00
15	1.00	1.00	1.00	1.00	2.00	4.00	4.00	3.00	2.00	1.00	1.00	42.00	2.00
16	1.00	2.00	1.00	2.00	2.00	4.00	2.00	4.00	1.00	3.00	2.00	46.00	3.00
17	2.00	2.00	1.00	1.00	1.00	1.00	4.00	3.00	4.00	2.00	2.00	32.00	1.00
18	1.00	1.00	1.00	2.00	2.00	4.00	1.00	4.00	1.00	1.00	1.00	45.00	3.00
19	2.00	2.00	1.00	1.00	2.00	2.00	3.00	2.00	3.00	4.00	2.00	27.00	1.00
20	2.00	2.00	2.00	1.00	1.00	3.00	4.00	2.00	3.00	1.00	2.00	30.00	2.00
21	1.00	1.00	1.00	2.00	2.00	5.00	5.00	4.00	2.00	2.00	1.00	48.00	3.00
22	2.00	2.00	1.00	2.00	2.00	3.00	1.00	4.00	3.00	4.00	2.00	37.00	3.00
23	1.00	2.00	2.00	1.00	1.00	4.00	4.00	3.00	2.00	1.00	1.00	31.00	2.00
24	1.00	1.00	1.00	1.00	2.00	4.00	4.00	3.00	2.00	1.00	1.00	35.00	2.00
25	1.00	2.00	1.00	2.00	2.00	4.00	2.00	4.00	1.00	3.00	2.00	35.00	3.00
26	2.00	2.00	1.00	1.00		1.00	4.00	3.00	4.00	2.00	2.00	25.00	1.00
27	1.00	1.00	1.00	2.00	2.00	4.00	1.00	4.00	1.00	1.00	1.00	37.00	3.00
28	2.00	2.00	1.00	1.00		2.00	3.00	2.00	3.00	4.00	2.00	27.00	1.00
29	2.00	2.00	2.00	1.00	1.00	3.00		2.00	3.00	1.00	2.00	30.00	2.00
30	2.00	1.00	2.00	2.00	2.00	4.00	2.00	4.00	2.00	1.00	2.00	39.00	2.00

In the data file, you can immediately see which cases have been selected. In Figure 5.9, respondents 2, 3, 4, 6 … (all women) have been literally scrapped. The selection can be undone by choosing the alternative "All cases" in the "Select Cases" menu. The deletions from the data file are then restored and all respondents are again involved in the analysis.

5.2 How do I calculate the frequencies of subgroups?

You can also select for age, which is done as follows:
> Data ▼
    > Select Cases ▼
      > If condition is satisfied ▼
      > If... ▼
      > age ▼
      > ▶ [to the window] ▼
      > type after 'age': GT 30
      > Continue ▼
    > OK ▼

With this procedure, use is only made of information from people older than 30 years old. Using the command GE instead of GT, the thirty-year-olds will also be included. You find the commands listed under "Function" in the Select cases window. Use can be made from such comparatives as:

LT = lesser than
GT = greater than
NE = not equal to
LE = lesser than or equal to
GE = greater than or equal to

5.3   How do I compare (sub)groups?

5.3.1   What is a clustered bar chart and a box plot?

clustered bar chart   A clustered bar chart is a bar chart in which the frequency distribution of a variable (e.g. *marital status*) is graphically represented with bars standing for two or more subgroups (e.g. men and women). In the case of the *marital status* variable, an indication is given in each category (single, with partner, with partner and children) of how many men and women are single, etc. The respondents are here clustered on the basis of their sex.

box plot   A box plot is a graphic representation of the distribution of a variable in terms of median and, additionally, on the basis of the middle 50% of the scores; that is the 2nd and 3rd quartiles.

quartile   Quartiles distribute the scores (which, for example, range from 3 to 15 on the happiness scale) in 4 segments each containing 25% of the respondents or cases. The first quartile (25th percentile) is the point on the scale under which a quarter of the scores lie. The scale point under which two quarters (= the half, 50%) of the scores lie is called the second quartile (50th percentile) and coincides with the median. Under the third quartile (or 75th percentile) lie 75% of the scores and above is of course 25%.

5.3.2   When do I use a clustered bar chart or a box plot?

A clustered bar chart can be used for nominal or discrete numerical variables, consisting preferably of not too many categories or values. It gives you some idea of the differences in a given characteristic (such as marital status) between groups (e.g. men and women).

A box plot can be used when you wish to visualize the differences between groups (e.g. men and women) for a given characteristic, such as happiness, measurable at the interval level, and when you wish to demonstrate the corresponding differences in standard deviation for the groups.

5.3.3   How do I make a clustered bar chart?

Although you could use 'select cases' to produce separate frequency printouts for men and women, there are other techniques in which you can highlight the differences between groups on a given variable. One of these is the clustered bar chart. To create this using SPSS, move the mouse indicator to "Graphs" on the menu bar and then select the roll-down menu entitled "Bar..."

Next go to "Clustered" and click it. You then have to use "Define" to indicate the variable for which the bar chart is to be made, and the variable on which clusters are to be based (see figure 5.10).
> Graphs ⊚
> > Bar...⊚
> > > Clustered ⊚
> > > Define ⊚

In the "Define Clustered Bar" menu, the *marital status* variable is the characteristic according to which men and women will be distinguished. You therefore select *marital status* and move this in the usual way to the field entitled "Category Axis":
> > marital status ⊚
> > ▶ [to the Category Axis field] ⊚

In the bar chart, you may then want to display the number of men and women to which each to the *marital status* categories applies; that is, the number of men and women who are single, living with partners or living with partners and children. The clusters are therefore defined according to *sex*.
> > sex ⊚
> > ▶ [to the Define Clusters by field] ⊚
> > OK ⊚

95

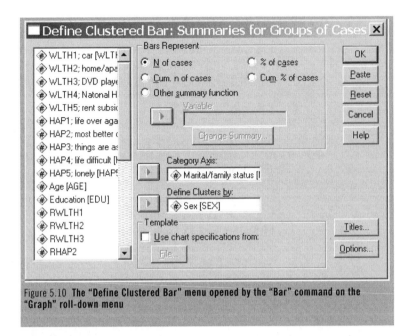

Figure 5.10 **The "Define Clustered Bar" menu opened by the "Bar" command on the "Graph" roll-down menu**

Review 5.2

How do I create a clustered bar chart?

> Graphs ⑦
> > Bar...⑦
> > > Clustered ⑦
> > > Define ⑦
> > > > VAR1 ⑦
> > > > ▶ [to the Category Axis field] ⑦
> > > > VAR2 ⑦
> > > > ▶ [to the Define Clusters by field] ⑦
> > > > OK ⑦

Figure 5.11 displays the results of this procedure.

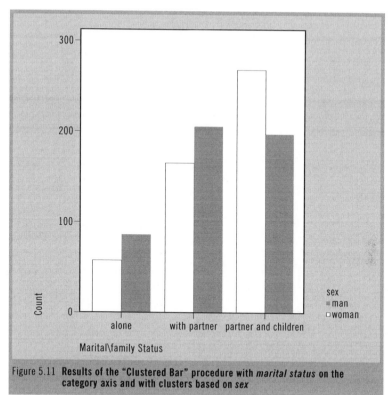

Figure 5.11 **Results of the "Clustered Bar" procedure with *marital status* on the category axis and with clusters based on *sex***

5.3.4   How do I construct a box plot?

For a variable measured at least on the interval level, there is another way to construct a clear diagram representing the distribution of its scores for one or more groups: the "box plot". This can also be found in the "Graphs" menu. If you would like to have separate representation of the distribution of the total happiness score for men and women, you then choose "Simple":

> Graphs ⊙
  > Boxplot ⊙
    > Simple ⊙

Next you go to "Define". You indicate here that the box plot is to be made for the *total happiness* variable and that separate box plots must be placed on the category axis for the values of the *sex* variable, hence separate ones for men and women. Clicking "OK" causes the procedure to be executed (see Figure 5.12):

> Define ⊙
  > THAP⊙
  > ▶ [to the Variable field] ⊙
  > sex ⊙
  > ▶ [to the Category Axis field] ⊙
  > OK ⊙

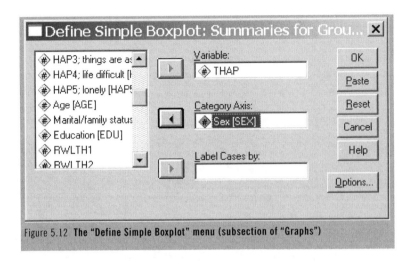

Figure 5.12 **The "Define Simple Boxplot" menu (subsection of "Graphs")**

The results of this procedure are shown in Figure 5.13.

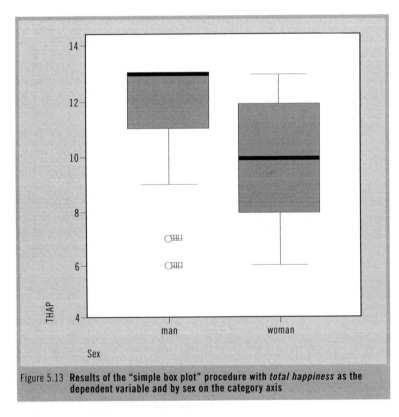

Figure 5.13 **Results of the "simple box plot" procedure with *total happiness* as the dependent variable and by sex on the category axis**

In Figure 5.13, you will see two boxes, one for men and another for women.

The bold line in the box plot represents the median, the middle-most value. This is 13 for the men and 10 for the women. Fifty

98

percent of the men have, consequently, a score of 13 or higher, 50% of the women a score of 10 or higher. Men are, as a group, happier than women. When the median is not located in the middle of the box, as is the case for men in the example, a negatively **skewed distribution** skewed distribution is being represented, in the sense that most scores are relatively high and there are only a few deviant (very) low scores. Only a few men are unhappy.

The box itself contains all the values that fall within the first and third quartiles, or between the 25th and 75th percentile scores. For men, this is therefore from 11 through 13; for women, it is lower, specifically from 8 through 12. The bottom of the box is represented by the first quartile; hence around 25% of women have a happiness score lower than 8. The top is formed by the third quartile: around 75% of the women in the sample have a happiness score of 12 or lower. The length of the box is the in- **interquartile range** terquartile range (IQR). In this case, the IQR for women is there- **IQR** fore 12 – 8 = 4. Fifty percent of the women have scores in the corresponding box and these therefore vary between 8 and 12. Above the box are the scores of 25% of the women.

For the men, there are a few anomalies at the bottom of the box plot, which are extreme scores and are called outliers. If scores **outliers** are more than 1.5 IQR removed from the bottom or top of the box, they are placed in a separate box plot. They are marked with an 'O' for 'outlier'. If they are exceptionally extreme, more than 3 IQR removed from the box, they are marked with an asterisk. Given that the IQR for the men is 2 (the box is located between the scores of 11 and 13), we regard scores lower than 11 – (1.5 * 2) = 8 as outliers. These are indicated by the 'O's besides the scores of 7 and 6 in the male group. Exceptionally extreme scores do not occur; there are therefore no asterisks in the box plot.

Additionally, there are lines above and below the box, which are **whiskers** called whiskers. They indicate the lowest possible values for scores that are not outliers. For the men, the bottom such value is 9, while for women it is 6, the lowest actual score among the women, which is well within the 1.5 IQR value delimiting the bottom of the scale for the female group (8 – [1.5 * 4] = 2) and which is not consequently an outlier. For the men, the top of the box is both the highest score and the median. There is consequently no top whisker. The advantage of using a box plot and the associated manner of describing the distribution relates to the method's reduced sensitivity to extreme values. In contrast, a mean is rather more sensitive to them, especially when small groups are involved. For that reason, use of a box plot for small groups is not very useful.

When you work with a clustered version of a box plot, you can even make comparisons on two levels. By making *total happiness* the variable, placing the *education* on the category axis and using *sex* as the cluster variable, we obtain the following printout (see Figure 5.14).

99

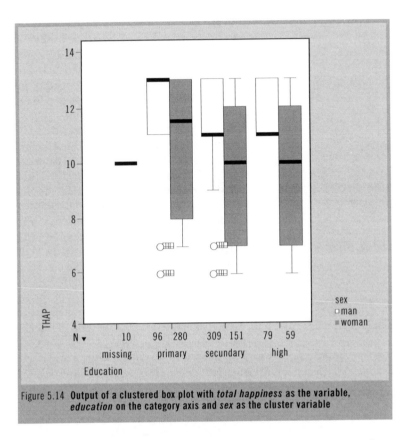

Figure 5.14 **Output of a clustered box plot with *total happiness* as the variable,** *education* **on the category axis and *sex* as the cluster variable**

In Figure 5.14, it immediately becomes clear that not only sex but also education reflect differences in feelings of happiness; happiest are men with the lowest education level, followed by women having the lowest level of education.

Review 5.3
How do I construct a box plot?

> Graphs ▼
  > Boxplot ▼
    > Simple ▼
    > Define ▼
      > VAR1 ▼
      > ▶ [to the Variable field] ▼
      > VAR2 ▼
      > ▶ [to the Category Axis field] ▼
      > OK ▼

5.4   How do I make variables comparable?

Variables are mostly distinct from each other because they are expressed in different units; for example, age (in years) and salary in euros. It may be the case that you would like to compare variables despite the differences in the manner of measurement. In our example, the variables *total wealth* and *total happiness* are scaled differently (respectively from 1 to 6 and 1 to 13). This makes comparison difficult. The sixth respondent has, for example, a score of 5 for wealth and 10 for happiness. Is his score for wealth then higher than for happiness? You can make it possible to compare the values of these variables by standardizing them, which means that you modify them, so they all have the same mean and same standard deviation. This procedure is called a linear transformation. When a variable is transformed into a score having a mean of zero and a dispersion of one, the transformation is regarded as one that produces a so-called z-score. A mean of 0 and a standard deviation of 1 are the specific features of a z-score. This linear transformation can be performed on SPSS by using the "Descriptives" sub-submenu found in the "Descriptive Statistics" submenu. Briefly stated, the procedure involves the following steps:

> Analyze ⊚
> > Descriptive Statistics
> > > Descriptives... ⊚

Next place a check in the box beside "Save standardized as valuables", hence:

> > > Save standardized as variables ⊚
> > > TWLTH ⊚
> > > ▶ [to the Variable(s) field] ⊚
> > > THAP ⊚
> > > ▶ [to the Variable(s) field] ⊚
> > > OK ⊚

In the data matrix, the new variables are found in the columns under the headings *ZTWLTH* and *ZTHAP*. Thus, SPSS automatically places a 'z' in front of the name, indicating that the variable's scores have been converted into z-scores. The z-scores of respondent 6 for *ZTWLTH* and *ZTHAP* are 0.13 and –0.38, respectively. The happiness score is negative, hence below the mean of 0, while the wealth score is positive, hence above the mean and therefore higher.

You can also check if the procedure was executed properly by requesting the mean and standard deviation for *ZTWLTH* and *ZTHAP*. You do that as follows:

> Analyze ⊚
> > Descriptive Statistics
> > > Descriptives... ⊚
> > > > ZTWLTH ⊚
> > > > ▶ [to the Variable(s) field] ⊚

101

> ZTHAP ⊙
> ► [to the Variable(s) field] ⊙
> OK ⊙

You then obtain the following output (Figure 5.15)

Descriptive Statistics

	N	Minimum	Maximum	Mean	Std.Deviation
Zscore (THAP)	984	−2.09188	0.91247	0.0000000	1.00000000
Zscore (TWLTH)	908	−1.50424	0.94645	0.0000000	1.00000000
Valid N (listwise)	902				

Figure 5.15 **Statistics for the standardized variables ZTWLTH and ZTHAP**

It is clear that the means for both variables are, in fact, 0 and the standard deviations are 1. The distributions remain, nevertheless, skewed: the distance from the maximum to the mean remains smaller than the distance from the minimum to the mean. The top of the distribution is, consequently, located somewhat to the right.

Key words

5.1.1
■ frequency distribution (Analyze > Descriptive statistics > Frequencies)
■ stem-and-leaf diagram
■ table
■ classes
■ valid percentage
■ cumulative percentage
■ mean
■ extreme scores
■ median
■ mode
■ standard deviation
■ variance
■ standard error of estimate
■ standard error (standard error of the mean)

5.1.2
■ graphs
■ circle diagram
■ pie chart
■ bar graph
■ bar chart
■ histogram
■ normal distribution (with normal curve)

5.1.3
■ valid cases
■ missing values
■ frequencies
■ value (code)
■ percent (percentage)
■ valid percent (percentage without missing values)
■ cum percent (cumulative percentage)
■ skewed distribution

5.2
■ select cases
■ if condition is satisfied

5.3
■ clustered bar chart (Bar Clustered)
■ box plot
■ quartile
■ interquartile range (IQR)
■ outliers
■ whisker

5.4
■ standardization of variables
■ linear transformation
■ z-score

102

6

How do I analyze my data when a comparative research question is involved?

Those who have some means
Think that the most important thing
in the world is love.
The poor know that it is money.

Gerald Brenan

1	2	3	4	5	6	7
How do I prepare myself to work with this Introduction?	How do I enter my data in the computer?	How do I modify or combine data?	How do I verify the homogeneity of the composite scores?	How do I analyse my data when a frequency research question is involved?	How do I analyse my data when a comparative research question is involved?	How do I analyse my data when a correlation research question is involved?

6.1	6.2	6.3
Comparative questions involving nominal test and grouping variables! Contingency tables and Chi-squares!	Comparative questions involving ordinal test variables and nominal grouping variables	Comparative questions involving an interval/ratio test variable and a nominal grouping variable

A comparative research question always involves the differences between two or more groups with respect to a particular variable. These groups are then distinguished on the basis of a given characteristic. Such a specific research question within the "Wealth and happiness" study asks if men are more or less content about their lives that they are leading than women are. The respondents are consequently divided according to "sex" into men and women. When a characteristic, such as sex in our example, is used to construct distinct categories, we call it the grouping variable. The variable according to which the men and women in our example are subsequently distinguished is called the test variable. If you would like to determine if there is a difference between the degree to which men and women are happy, sex remains the grouping variable and happiness becomes the test variable. In the preceding chapter, we saw that there are various techniques to make the differences between groups clearly apparent. You can, for example, use "Select Cases" to compute the individual counts for different marital status categories or the separate total happiness scores for men and women. If mean total happiness scores are calculated separately for men and women, it appears that the mean score for men (mean = 11.50) is higher than that of women (mean = 10.27). If you are working with a population (all the people about whom you wish to make statements are then included in your study), you can say that the men on average are happier than women. If you are using a sample (only a portion of the population is included in the study), you cannot then simply make such a claim. There is, after all, always the possibility that the results were due to chance. On one occasion the sample mean may be higher and on the next occasion lower. It is therefore necessary to test how great the possibility is that the different means are attributable to chance or, in other words, if a difference is significant. How great, for example, is the chance that the difference between the means scores (the difference

Prior knowledge
We assume that:
- you can start SPSS for Windows (1.4);
- you can retrieve a data file (2.6);
- you can distinguish among the various levels of measurement (1.2.2);
- you know what discrete and continuous variables are (1.2.2);
- you know what is involved in executing the SPSS select cases procedure (5.2);
- you know what is meant by mean and standard deviation, as well as how to compute them (5.1.1);
- you know what a unit of research (unit of analysis), population and sample are (1.3.3);
- you understand the expressions: probability of error (p value) and degrees of freedom (= df) (1.3);
- you know what we mean by significance (1.3);
- you can use SPSS to recode (3.1);
- you know the difference between one and two-tailed tests (1.3);
- you know what a normal distribution is (1.3);
- you know what a standard error is (1.3).

104

between 11.50 and 10.27) is due to coincidence, when it is a sample that is actually being measured? Significance is determined by:

- the size of the observed differences; the greater the differences, the smaller the possibility that they are a result of chance;
- the size of the sample; the larger the samples, the smaller the possibility that a difference is a result of chance;
- the variation (differences) within the groups; the smaller the variation within the samples, the smaller the possibility that a difference between groups results from chance;
- the chosen level of significance; the more certain you wish to be that the difference is not due to chance – in which case you make the alpha as small as possible – the smaller the chance that a difference will be significant.

To determine the extent to which a difference may be due to chance, use is made of statistical tests that, for the most part, take the above-mentioned factors into account. The particular statistical test to be employed is determined by the measurement level of the test variable. In Section 6.1, we deal with the statistical tests to be used when both the test variable and the grouping variable are measured at the nominal level. Section 6.2 is concerned with the tests employed when the grouping variable is at least at the nominal level and the test variable at the ordinal level. The tests used when the grouping variable is at least measured at the nominal level and the test variable at the interval or ratio level are discussed in Section 6.3. When choosing an adequate statistical test to establish if differences in certain variables exist between two or more groups, it is especially necessary to determine if paired (dependent) or independent samples are involved. By independent samples, we mean that pure chance determines if Respondent 1 is placed in group A and Respondent 2 in group B. When Respondent 1 in group A and Respondent 2 in group B come from the same family,

Introduction

the study is then employing paired or dependent samples. Such

may be the case in a study that investigates, for example,

differences between the oldest and youngest children in a family.

When you make two separate measurements of the same person,

as is the case when you measure a person's blood pressure before

and after the administration of a drug, paired sampling

techniques are then clearly being used.

Moreover, all tests require that samples are aselect, which

means that the choice of population members included in the

sample must be a completely random one. Every individual must

have an equal chance of being included in the sample.

We will not only indicate how you can test if a difference is

significant but also how you can determine the extent of the

difference. With a large sample, an extremely small difference

can, for example, appear to be significant. Moreover, such a

difference may indeed be significant but not relevant, since it

demonstrates so little. There are various ways to discern the

effect of the grouping variable on the test variable (effect size);

these will be discussed for each test.

6.1 Comparative questions involving nominal test and grouping variables? Contingency tables and Chi-squares!

6.1.1 What is a contingency table and Chi-square?

contingency table

A contingency table summarizes the scores for a variable X in combination with the scores for a second variable Y. It consequently displays the joint frequency distribution of two characteristics measured on the same persons (cases). One variable is listed down the left side of the table and labels the values in the rows of the tables. The other variable is recorded across the top of the table and indicates the values in the columns. The cells or boxes of the tables contain the number of respondents or cases having the various combinations of the two characteristics (men + single, women + single, etc.).

Table 6.1 **The number of (non-)single individuals grouped according to sex (absolute numbers)**		
	Man	Woman
Single	61	89
Non-single	438	408

The Chi-square test can be used to investigate if a significant difference exists between two or more groups for one characteristic measured at the nominal level. Each cell of the contingency table compares the number of observed cases with the number that might be expected on the basis of chance (expected cell counts). The sample in the "Wealth and happiness" study contains an equal number of men and women. If 150 of them are single, it would be expected on the basis of chance that these 150 singles would be comprised of approximately 75 men and about the same amount of women. You can calculate this exactly by dividing the product of both relevant marginal totals by the total sum (see Table 6.2); consequently, for the first cell: (150 * 499) / 996 = 75. We are, however, presented in the sample with 61 single men and 89 single women.

Table 6.2 **The number of (non-)singles grouped according to sex, along with the expected cell counts (in parentheses) and the marginal totals.**			
	Man	Woman	Total
Single	61 (75)	89 (75)	150
Non-single	438 (424)	408 (422)	846
Total	499	497	996

This result deviates from the one that you would expect on the basis of chance. Is this deviation great enough to be significant and not merely dependent on chance? An affirmative answer to the above question would lead you to conclude that there were more single women in the population than single men. The correctness of the answer can be tested using the Chi-square test (Chi^2), a calculation involving the difference between the observed counts and the expected counts on the basis of chance.

The formula appears as follows:

$$x^2 = \sum \frac{(O - E)^2}{E}$$

in which: O = the observed cell count
E = the expected cell count

107

6.1 Comparative questions involving nominal test and grouping variables? Contingency tables and Chi-squares!

In the example, the equation would therefore appear:

$$x^2 = \frac{(61-75)^2}{75} + \frac{(89-75)^2}{75} + \frac{(438-424)^2}{424} + \frac{(408-422)^2}{422} = 61,5$$

The greater the Chi-square, the smaller the probability that the deviation results from chance (the p value). SPSS does not only indicate the Chi-square but also the corresponding p value, which is not only dependent on the size of the Chi-square but, additionally, on the number of degrees of freedom (df). In a contingency table, these are determined by multiplying the number of columns minus 1 by the number of rows minus 1:

degrees of freedom
df

$$df = (k-1)(r-1)$$

in which: c = the number of columns
r = the number of rows

In the example, the equation would therefore become:

$$df = (2-1)(2-1) = 1$$

The fewer the number of columns and rows, the smaller the number of degrees of freedom and the more likely that the Chi-square is significant. The p value corresponding to a Chi-square of 6.15 and a df = 1 is p < 0.05, which SPSS automatically displays. With an alpha of 0.05, this difference is therefore significant.

6.1.2 When do I use a contingency table or Chi-square?

A contingency table can be used to present data measured at a nominal level (sex, religion) and ordinally discrete data (for example, education divided into the categories of lower secondary, upper secondary and higher).

For use to be made of the Chi-square:
- any given *expected* cell frequency may not be smaller than 1;
- a minimum of 80% of the *expected* cell counts must have values greater than 5;
- the variables must not have too many categories (otherwise the table becomes incomprehensible and the first two conditions are also likely not satisfied).

The SPSS output from a Chi-square test indicates if the first two conditions have been met. When the table involved has only two rows and two columns, and the above conditions of the Chi-square test are not satisfied, SPSS automatically makes use of

108 **Fisher's exact test** Fisher's exact test, in which such conditions do not apply.

6.1.3 How do I compose a contingency table and to calculate a Chi- square?

You first need to retrieve a data file in the usual manner. In this case, we are making use of 'data5.sav'. Select "Analyze" in the main menu and subsequently "Descriptive Statistics". Then move the mouse indicator to "Crosstabs" in the roll-down menu that appears:

> Analyze ⊙
> > Descriptive Statistics
> > > Crosstabs... ⊙

These commands are used to indicate that, in your analysis, you wish to employ one method of descriptive statistics: you would like to have SPSS compose a contingency table. In addition, you need to indicate how the resulting contingency table will appear and which variables need to be included in it. First of all, you **row variable** must define the row variable ("Row(s)" = horizontal). Although it **"Row(s)"** is, of course, not possible in the example to speak of a causal relationship between the *sex* and *marital status* variables, clearly *sex* is the grouping variable and *marital status* is the test variable. After all, marital status can never have an influence on the respondent's sex, whereas the inverse is theoretically possible. Conventionally, the test variable is presented as the row variable **column variable** in the contingency table and the grouping as the column variable **("Column(s)"** ("Column(s)" = vertical).

See Figure 6.1.

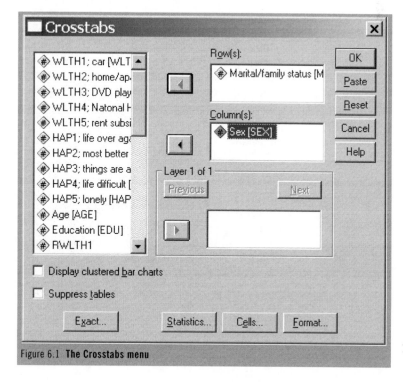

Figure 6.1 **The Crosstabs menu**

109

Place the test variable (in this case, *marital status*) in the box under "Row(s)" and the grouping variable (in this case, *sex*) in the field under "Column(s)":

> marital status ▼
> ▶ [to the Row(s) field] ▼
> sex ▼
> ▶ [to the Column(s) field] ▼
> OK ▼

The standard output from this procedure is a contingency table; see Figure 6.2.

Marital status * sex Cross tabulation

Count

		Sex		
		Man	Women	Total
Marital status	single	61	89	150
	with partner	168	208	376
	with partner and children	270	200	470
Total		499	497	996

Figure 6.2 The standard output for the "Crosstabs" procedure in which *marital status* is the test variable and *sex* is the grouping variable.

The standard contingency table provides an indication of the number of cases or respondents (the absolute count). This requires nothing more than a count to be made of the various cases for each category of the variables involved. For this reason, the term "Count" appears in the top left corner of the contingency table. From the row totals, it can be seen that 150 respondents are single, 376 live with a partner and 470 respondents with a partner and children. The column totals indicate that there are 499 men and 497 women in the sample. There are 4 cases involving a missing value. The uppermost row shows that 61 men and 89 women are single, etc.

Count

Cells

row percentages
column percentages
total percentages
Expected

Supplementary information can be retrieved by using the "Cells" command (at the bottom of the "Crosstabs" menu; see Figure 6.1). You can click "Row" to obtain the row percentages, "Column" to get the column percentages and "Total" for the table or total percentages (see Figure 6.3). You can even request the expected count ("Expected") for each cell. Using the output presented in Figure 6.4, we will illustrate what is meant by all of the above. You first must click "Analyze", "Descriptive Statistics" and the "Crosstabs" command, followed by:

> Cells... ▼

110

The "Crosstabs: Cell Display" window then appears (Figure 6.3). You are then able to simultaneously check the options "Row", "Column", "Total", "Expected" and "Observed":

> Row ⊙
　　> Column ⊙
　　> Total ⊙
　　> Expected ⊙
　　> Observed ⊙
　　> Continue ⊙
> OK ⊙

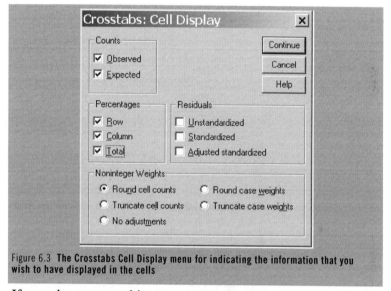

Figure 6.3 **The Crosstabs Cell Display menu for indicating the information that you wish to have displayed in the cells**

If you choose everything at once, each cell in the table will display the row, column and total percentages, along with the observed and expected counts (see Figure 6.4).

111

Marital status * sex Cross tabulation

				Sex		Total
				Man	Women	
Marial status	Single		Count	61	89	150
			Expected Count	75.2	74.8	150.0
			% within marital status	40.7%	59.3%	100.0%
			% within sex	12.2%	17.9%	15.1%
			% of Total	6.1%	8.9%	15.1%
	With partner		Count	168	208	376
			Expected Count	188.4	187.6	376.0
			% within marital status	44.7%	55.3%	100.0%
			% within sex	33.7%	41.9%	37.8%
			% of Total	16.9%	20.9%	37.8%
	With partner and children		Count	270	200	470
			Expected Count	235.5	234.5	470.0
			% within marital status	57.4%	42.6%	100.0%
			% within sex	54.1%	40.2%	47.2%
			% of Total	27.1%	20.1%	47.2%
Total			Count	499	497	996
			Expected Count	499.0	497.0	996.0
			% within marital status	50.1%	49.9%	100.0%
			% within sex	100.0%	100.0%	100.0%
			% of Total	50.1%	49.9%	100.0%

Figure 6.4 **The output from the "Crosstabs" procedure in which "Cells" has been used to request the row, column and total percentages, along with the expected cell count**

Of course, the marginal totals of the contingency table (row and column totals) are the same. This output does, however, provide a great deal more information. Each cell of the contingency table contains 5 figures. The left column of the table indicates what these figures represent:

Count
- The figure at the top is still the absolute number (Count). For example, there are 61 men who live alone.

Expected count
- The number of single men to be expected purely as a result of chance (Expected count is 75.2. The actual number (61) can be seen to deviate somewhat from that amount. In addition, differences between observed and expected counts can also be seen to occur in other cells.
- If we consider all respondents who live alone (n = 150) to be 100% (of the singles in the study), 40.7% of this category of

row percentage
people (row percentage) are comprised of men; in Figure 6.4, this is indicated as the '% within marital status'.
- If we consider all men (n = 499) to be 100%, 12.2% of this cate-

column percentage
gory of people (column percentage) are comprised of singles; in Figure 6.4, this is indicated as the '% within sex'.
- If we consider all respondents to be 100%, 6.1% of them are

total percentage
shown to be men and single (total percentage); the 61 single men constitute 6.1% of the total group, and this figure is reported as the '% of Total'.

We have shown that a difference exists between observed and expected counts. We can learn if this difference is so great that it is, in addition, significant by having SPSS calculate the Chi-square.

112

In this way, differences between the observed and the expected are tested. The greater the differences, the greater the Chi-square and the greater the probability that a significant difference is involved. You can retrieve statistical data from the contingency table by clicking the "Statistics" button in the "Crosstabs" menu (Figure 6.1). In a contingency table involving a sample, you are, of course, interested in performing the Chi-square test. You therefore need to execute the following commands in the "Crosstabs" window:

> Statistics... ⊙
> > Chi-Square [check] ⊙
> > Continue ⊙
> OK ⊙

See Figure 6.5.

Chi-Square Tests

	Value	df	Asymp. Sig. (2-sided)
Pearson Chi-square	19.904[a]	2	0.000
Likelihood Ratio	19.981	2	0.000
Linear-by-Linear Association	18.309	1	0.000
N of Valid Cases	996		

a. 0 cells (0.0%) have expected count less than 5. The minimum expected count is 74.85.

Figure 6.5 **Output of the "Crosstabs" procedure, along with the "Statistics" procedure "Chi-square"**

6.1.4   How should I read the results of a Chi-square test?

The difference in the marital status of men and women is clearly significant. Chi-square (otherwise known as Pearson's Chi-Square) is 19.90, and the probability that the observed marital status difference between men and women at two degrees of freedom (df = 2) results from chance is 0.000. The number of degrees of freedom is equal to the product of the number of columns (man, woman) minus one and the number of rows (single, with partner, with partner and children) minus one. Therefore, in this case $(2 - 1) * (3 - 1) = 2$. The probability that there is no difference between men and women in the population given this Chi-square (19.90) and two degrees of freedom (df) is smaller than one-tenth of a percent. You can conclude with a certainty greater than 95% (in this case even greater that 99.9%) that a significant difference exists in the marital status of the men and women. SPSS also indicates in the output if you have satisfied the conditions of a Chi-square test (see Figure 6.5). As it happens, all expected cell counts were greater than 5. If you do not meet the conditions because some cells are insufficiently full, you can then sometimes resolve this problem by using the Recode procedure (Section 3.1) to group some of the values. The new groupings must, of course, be meaningful insofar as content is concerned. You could, for example, make one category of people living to- **113** gether with or without children: 'living with others'. However, this is not necessary for the "Wealth and happiness" data file. The 'Likelihood Ratio' (see Figure 6.5) is comparable to the Chisquare test and furnishes nearly the same results for large(r) numbers of respondents.

Once it has been established that a significant difference exists between men and women insofar as marital status is concerned, the next question to be asked involves the extent to which marital status can be explained by sex. To answer this question, Chisquare must be transformed into the measure of association known as Cramér's V, which has a value between 0 and 1. In this regard, 0 indicates that no association exists and 1 that there is a perfect association. We will explain how you compute Cramér's V in Section 7.1.

6.1.5 How should I report the results of a contingency table or a Chi- square test?

It is unusual for the output, such as that shown in Figure 6.2, to be included in its entirety in a research report. You should make a selection from it on the basis of your research question. If this question asks: 'Is there a difference between men and women insofar as marital status is concerned?' then the table would appear different from the one corresponding to the question: 'Is there a sex-based difference between those who live together with others and those who do not?'

A rule of thumb is: if you compare horizontally, list percentages vertically, and if you make vertical comparisons, list percentages horizontally.

Table 6.3 **Summary of the differences in marital status between men and women**

Marital status	Man		Woman	
	N	%	N	%
Single	61	12	89	18
With partner	168	34	208	42
Partner and child(ren)	270	54	200	40
Total	499	100	497	100

If you wish to compare men and women in your sample in terms of their marital status (that is, make a horizontal comparison), you should then list percentages vertically and, consequently, present the percentages in columns; see Table 6.3 for an example. If, conversely, you wish to examine if respondents living with others (partner and children) and those living alone may differ in terms of sex (that is, make a vertical comparison), you will then have to list percentages horizontally in rows.

Table 6.4 **Summary of sex-based differences according to marital status**

marital status	Man		Vrouw		Totaal	
	N	%	N	%	N	%
Single	61	41	89	59	150	100
With partner	168	45	208	55	376	100
Partner and child(ren)	270	57	200	43	470	100

Note that tables 6.3 and 6.4 are, with the exception of the percentages, identical to each other.
The results of the testing can be described as follows:

"A significant difference has been demonstrated to exist between men and women insofar as marital status is concerned. Men live together with both a partner and children more frequently than women do. Women, conversely, live alone or only together with a partner (without children) more frequently than men do (Chi² = 19.9; df = 2; p < 0.001)."

To illustrate this result, you can include the contingency table in your research report (Table 6.3).

Review 6.1
How do I compute differences in a study involving a nominal test variable and a nominal grouping variable?

> Analyze ▾
> > Descriptive Statistics
> > > Crosstabs... ▾
> > > > VAR1 ▾
> > > > ▶ [to the Row(s) field] ▾
> > > > VAR2 ▾
> > > > ▶ [to the Column(s) field] ▾
> > > > Cells... ▾
> > > > > Expected ▾
> > > > > Row ▾
> > > > > Column ▾
> > > > > Total ▾
> > > > > Continue ▾
> > > > Statistics... ▾
> > > > > Chi-Square [check] ▾
> > > > > Continue ▾
> > > > OK ▾

6.2 Comparative questions involving ordinal test variables and a nominal grouping variable?

6.2.1 Comparative questions involving two independent samples? Mann-Whitney U test!

6.2.1.1 What is the Mann-Whitney U test?

Mann-Whitney U

The Mann-Whitney U test can be used to investigate if a difference between two independent samples concerning an ordinal variable Y is due to chance. Table 6.5 contains the happiness scores for a sample of 5 male and 5 female employees of the 'Labour' company. It is clear that the scores of the five women are, on average, higher than those of the men. It remains to be seen if this is a consequence of chance or if it is a systematic difference. With the Mann- Whitney U test, you can test what the probability is of finding such a difference merely as a result of chance. To this end, all respondents (men and women) are grouped together and each allotted a rank score on the basis of

rank score

their score on the Y variable (happiness). Accordingly, the lowest score (8) receives a value of 1, the next three employees, all having a score of 9, are then each awarded the rank score 3 (the mean of the rank scores 2, 3 and 4). Subsequently, the rank scores of both groups (men and women) are separately totalled. You can see that the total of the rank scores for women (31.5) is higher than the one for men (23.5). The extent to which the rank scores

value 'U'

are different from each other is expressed as the value 'U'. Whether or not this difference is significant depends on the size of the sample and the extent of the difference. These considerations are taken into account in the formula for U, which is calculated for both samples separately:

$$U_1 = n_1 n_2 + \frac{n_1(n_1+1)}{2} - R_1$$

$$U_2 = n_1 n_2 + \frac{n_2(n_2+1)}{2} - R_2$$

The smaller of the two U values is used to establish if a finding is significant.
For the example, the U values are:

$$U_1 = 5 * 5 + \frac{5(5+1)}{2} - 31.5 = 8.5$$

$$U_2 = 5 * 5 + \frac{5(5+1)}{2} - 23.5 = 16.5$$

The smaller U value of 8.5 has, in this sample, a corresponding p value of about 0.40, which is much higher than an alpha of 0.05 and therefore indicates that a significant difference is not in-

116

volved. When there are respondents or cases sharing the same (rank) score (a tie), such as is the case of the three male employees who all have a score of 9, a correction is then performed by SPSS. In practice, the corrected values hardly deviate from the uncorrected ones.

Table 6.5 **Summary of the happiness scores for a sample of 5 male and 5 female employees of the 'Labour' company**

Female employees		Male employees	
Happiness	Rank score	Happiness	Rank score
14	10	13	9
12	8	9	3
8	1	10	5.5
11	7	9	3
10	5.5	9	3
Total	$R_1 = 31.5$		$R_2 = 23.5$

6.2.1.2 When do I use the Mann-Whitney U test?

A Mann-Whitney U test can be used when:
- the grouping variable comprises two independent samples;
- the test variable has been measured at least at the ordinal level.

The Mann-Whitney U test is often used when small samples (< 25) are involved, even when the relevant variables are measured at the interval or ratio level and the use of the t test is indicated (see Section 6.4.1). In small samples, the t test is more sensitive to extreme scores than the Mann-Whitney U test is.

When you wish to compare three or more groups, for example, people with a lower secondary, upper secondary and higher education, the Mann-Whitney U test cannot, however, be used; you must use the Kruskal-Wallis test instead.

6.2.1.3 How do I perform a Mann-Whitney U test?

Let us consider another specific research question that asks: 'Is there a difference between men and women in the extent that they are happy about the life that they are leading?' In such a case, the issue involves two variables of which the grouping variable is measured at the nominal level (i.e. sex) and the test variable at the ordinal level (i.e. happiness; see Section 1.2). To obtain an answer to the stated research question, you should select "Analyze" in the main menu and then "Nonparametric Tests" in the roll-down menu that appears. In an investigation of whether or not men are on average happier than women, the two categories (men and women) may be characterized as independent 'samples'. You can therefore make use of the Mann-Whitney U test:

117

> Analyze ⊙
  > Nonparametric Tests
    > 2 Independent Samples ⊙

See Figure 6.6.

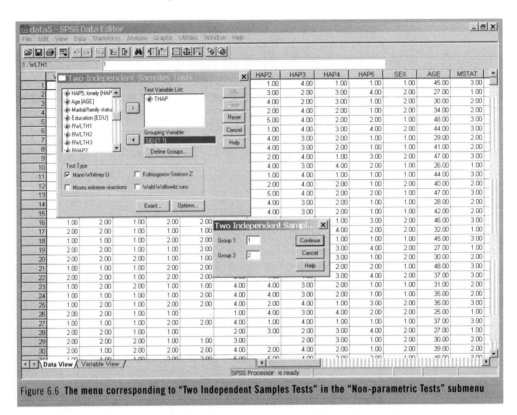

Figure 6.6 **The menu corresponding to "Two Independent Samples Tests" in the "Non-parametric Tests" submenu**

Two Independent Samples Tests

You will then arrive in the dialogue window "Two Independent Samples Tests". Mark the test variable in the variable list (in this case *THAP*) and move it to the frame under the heading "Test Variable List". Mark the grouping variable in the same manner (in our example, *sex*) and move it to the frame under the "Grouping Variable" heading:

> THAP⊙
> ▶ [move to the Test Variable List field] ⊙
> sex ⊙
> ▶ [to the Grouping Variable field] ⊙

As soon as you have placed the grouping variable *sex* in the "Grouping Variable" field, two question marks appear behind the variable name; hence in this case: 'sex(??)'. Using the "Define Groups" button, you can indicate which categories or groups are to be distinguished. Type 1 in the field next to "Group 1" and 2 in

118

the field next to "Group 2". Sex is consequently numerically coded with a number 1 for men and a number 2 for women:

> Define Groups... ⊙
> Group 1 ⊙
> type: 1
> Group 2 ⊙
> type: 2
> Continue ⊙

Once you have clicked "Continue", you will then see the indicated codes (1 & 2) behind the word 'sex' instead of the two question marks. If a check appears before Mann-Whitney U in the "Test Type" field, the procedure can then be executed. If this mark does not appear, the following steps should then be taken:

> Test Type: Mann-Whitney U ⊙
> OK ⊙

The output from this procedure is presented in Figure 6.7:

Ranks

	Sex	N	Mean Rank	Sum of Ranks
THAP	man	484	571.97	276831.49
	Woman	500	415.58	207788.50
	Total	984		

Test Statistics[a]

	THAP
Mann-Whitney U	82538.500
Wilcoxon W	207788.500
Z	-8.906
Asymp.Sig. (2-tailed)	0.000

a. Grouping Variable: sex

Figure 6.7 **The output from the "Two-Independent Samples" procedure in the "Nonparametric Tests" screen**

6.2.1.4 How should I read the results of a Mann-Whitney U test?

The output, which consists of two tables, indicates the number of respondents (n) for both groups. There are 984 respondents involved in this analysis: 484 men and 500 women. The mean THAP rank is 571.97 for group 1 (men) and 415.58 for group 2 (women). This makes it clear that the women score lower in happiness, insofar as the mean rank is concerned. In addition to the mean rank, the sum of ranks is also indicated. This test is based on the smallest sum, which is, in this case, 207788.50. The Test Statistics table lists the statistic U and the z value derived from U. The U value is calculated on the basis of the smallest sum of the ranks, along with the sample sizes. This U value can, with the aid of a formula, be converted into a z value. The values are corrected for the number of ties. We first note the U value, which, in this case,

mean rank
sum of ranks

U value

119

is 82538.500 and which has a very high corresponding z value of –8.90. The (two-tailed) probability of error corresponding to this U value and this number of respondents is 0.000. Consequently, the probability that this difference is a result of chance is smaller than 1/1000 and the difference is therefore significant. If, on the basis of certain considerations, you might expect that women are unhappier than men (or vice verse), then you will need to conduct a one-tailed test. It is difficult to establish the effect size, as the test variable is, in this case, measured at the ordinal level. For that reason you can draw a conclusion in terms of more or less happy without being able to say anything about how much more and less that it may be.

Wilcoxon rank sum test
Instead of the Mann-Whitney test, the Wilcoxon rank sum test is then used, in which the statistical quantity U is not calculated but a z value. SSPS also indicates the z value and automatically includes it in the output. This Wilcoxon test should not be confused with the Wilcoxon Signed-ranks test, which we will discuss in 6.2.3.

6.2.1.5 How should I report the results of a Mann-Whitney U test?

Performing the Mann-Whitney U test on the above reveals that the resulting probability of error, found to be 0.000, is much smaller than 0.05 (the usual alpha or unreliability of the test). It can therefore be concluded that a significant difference exists between men and women in the extent to which they say that they are happy.

In a research report, such a finding can be formulated as follows:

"It has been demonstrated that men (mean rank score 572) are happier than women (mean rank score 416). This difference is significant (U = 82538.5; p < 0.001)."

To illustrate your conclusion, you could include the box plot shown in Figure 5.13.

▬ **Review 6.2**
How do I compute the differences in a study involving two independent samples in which there is an ordinal test variable and a nominal grouping variable?

> Analyze ⊙
   > Nonparametric Tests
      > 2 Independent Samples ⊙
         > VARy ⊙
         > ▶ [move to the Test Variable List field] ⊙
         > VARx ⊙
         > ▶ [to the Grouping Variable field] ⊙

> Define Groups... ⊙
  > Group 1: type for example: 1
  > Group 2: type for example: 2
  > Continue ⊙
> Test Type: Mann-Whitney U ⊙
> OK ⊙

6.2.2 Comparative questions involving three or more independent samples? Kruskal-Wallis test!

6.2.2.1 What is the Kruskal-Wallis test?

Kruskal-Wallis test

The Kruskal-Wallis test is based on the same principle as the Mann-Whitney U test. The groups of respondents or cases (for example, single, with partner, and with partner and children) are mixed together to form a total group. All respondents are then ranked and are therefore given a rank score on the test variable (for example, THAP). Ties, when different respondents receive the same (rank) score, are corrected. The relevant rank scores are then totalled for each group and the mean rank score calculated. The differences between them provide the basis on which the statistic U is calculated and subsequently converted into a Chi-square value.

6.2.2.2 When do I use the Kruskal-Wallis test?

You can use the Kruskal-Wallis test when:
- more than two independent samples are involved;
- the test variable is at least measured on the ordinal level.

6.2.2.3 How do I perform a Kruskal-Wallis test?

When more than two groups are involved in your study, such as those having lower secondary, upper secondary and higher education, the Kruskal- Wallis test can be used. To do this, you select "Analyze" in the main menu and subsequently "Nonparametric Tests". We are investigating if a difference in happiness exists among people having different marital status: single, living with a partner, or living with a partner and children. The three categories (single, with partner, and with partner and children) can be characterized as three independent samples. You can therefore make use of the Kruskal-Wallis test, which you will find under the heading "K Independent Samples (see Figure 6.8).
> Analyze ⊙
  > Nonparametric Tests
    > K Independent Samples ⊙

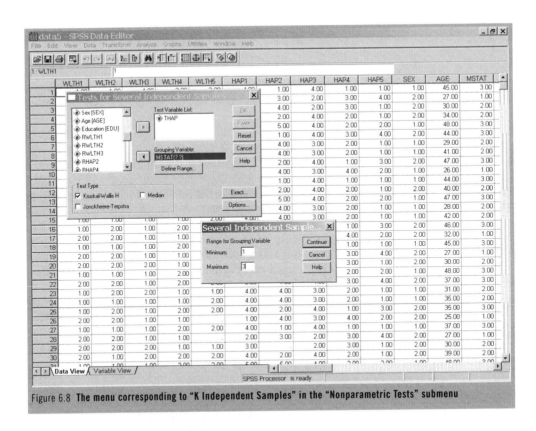

Figure 6.8 **The menu corresponding to "K Independent Samples" in the "Nonparametric Tests" submenu**

You will then arrive in the dialogue window entitled "TESTS FOR SEVERAL INDEPENDENT SAMPLES". Mark the test variable in the variable list (in this case, *THAP*) and move it to the frame under the heading "Test Variable List". Mark the grouping variable in the same manner (*marital status* in our example) and move it to the frame having "Grouping Variable" as a heading:

> THAP⊙
  > ► [move to the Test Variable List field] ⊙
  > marital status ⊙
  > ► [to the Grouping Variable field] ⊙

As soon as you have placed the grouping variable *marital status* in the "Grouping Variable" field, two question marks appear after the variable name; hence in this case: 'marital status(??)'. Using the "Define Range" button, you can indicate the categories or groups that are to be distinguished. You then arrive in a submenu (see Figure 6.8). Type a 1 in the field next to "Minimum" and a 3 in the frame next to "Maximum". Marital status is, of course, coded with the code 1 for 'single', 2 for 'partner' and 3 for 'partner + children'. The scores consequently range from a minimum of 1 to a maximum of 3.

122

> Define Range... ▼
  > Minimum: ▼
    type for example: 1
    Maximum ▼
    type for example: 3
  > Continue ▼

Once you have clicked "Continue", you will see that the two question marks behind "marital status" are replaced by the entered codes. The command can now be executed.
  > OK ▼

The output from this procedure is presented in Figure 6.9.

Ranks

	marital status	N	Mean Rank
THAP	single	147	76.29
	with partner	376	406.83
	with partner and children	457	692.58
	Total	980	

Test Statistics[a,b]

	THAP
Chi-Square	618.521
df	2
Asymp. Sig.	0.000

a. Kruskal Wallis Test

b. Grouping Variable: marital status

Figure 6.9 **The output from the "K Independent Samples" procedure in the "Nonparametric Tests" submenu**

6.2.2.4 How should I read the results of the Kruskal-Wallis test?

The output of the Kruskal-Wallis test consists of two tables: Ranks and Test Statistics. The Ranks table lists the number of respondents or cases (N) and the mean rank score (Mean Rank). There are, for example, 147 of the 980 respondents who are single. Their mean rank score on the *THAP* variable is 76.29. The means rank score for people with a partner is 406.83, and that for the individuals with a partner and children is still higher: 692.58. The test results are presented in the Test Statistics table. Chi^2 is used to test the differences among the mean rank scores. In our case, this is corrected for ties, 618.5; the number of degrees of freedom (df) is 2 (i.e., k − 1, in this example 2 groups − 1 = 2). The p value or probability of error is 0.000. It is therefore improbable that the difference in the happiness scores of the three groups is attributable to chance.

It is difficult to establish the effect size, as the test variable is, in this case, measured at the ordinal level and you can only draw a

123

conclusion in terms of more or less happy without being able to say anything about how much more and less that it may be.

6.2.2.5 How should I interpret and report the results of the Kruskal-Wallis test?

The differences in mean rank score on the *THAP* variable are both substantial and significant. Singles have the lowest score and respondents who live with a partner and children the highest. Respondents with a partner alone occupy an intermediate position.

In your research report, you could report this result as follows:

"The study demonstrates that a difference in the experience of happiness exists among people with different marital statuses. Respondents living with a partner and children are the most content about the life that they are leading or, in other words, the happiest. The next happiest is the group of respondents who live with a partner, while singles are lowest on the happiness scale. The differences in happiness among these groups are significant (Kruskal-Wallis test $Chi^2 = 618.52$; df = 2; p < 0.001). The difference between singles and both the other groups is, in a negative sense, greater than the difference between respondents with partners and respondents with partners and children. The mean rank score for singles (n = 147) is 76.29; for people with partners (n = 376) it is 406.83, and 692.58 for people with partners and children (n = 457)."

Review 6.3

How do I compute the differences in a study involving three or more independent samples in which there is an ordinal test variable and a nominal grouping variable?

> Analyze ▼
> > Nonparametric Tests
> > > K Independent Samples ▼
> > > > VARy ▼
> > > > ▶ [move to the Test Variable List field] ▼
> > > > VARx ▼
> > > > ▶ [to the Grouping Variable field] ▼
> > > > Define Range... ▼
> > > > > Minimum: type for example: 1
> > > > > Maximum: type for example: 3
> > > Continue ▼
> > > OK ▼

6.2.3 Comparative questions involving two paired samples? Wilcoxon Signed-ranks test!

6.2.3.1 What is the Wilcoxon Signed-ranks test?

With the Wilcoxon Signed-ranks test, you can discover if a difference on an ordinal variable Y between two paired samples can be attributed to chance. Y is, for example, happiness and the grouping variable is X; for example, two positions in the order of children born to a family, the categories being the 'oldest child' and the 'youngest child'. The issue therefore involves pairs of respondents, in this case two children who always come from the same family. First, the difference between the individuals in each pair of Y is calculated. In the example, this is the difference of the happiness variable. The absolute differences are then ordered from lowest to highest. Next, the rank scores of all negative differences are summed and similarly all the positive differences. The mean positive rank score is finally compared with the mean negative rank score. See the example in Table 6.6.

Table 6.6 **Happiness scores of the oldest and youngest children from the same family (range of scores = 3 through 15)**

Pair	Happiness score of oldest child	Happiness score of youngest child	Difference	Absolute difference	Rank
A	5	12	−7	7	4
B	4	10	−6	6	3
C	7	6	+1	1	1
D	10	8	+2	2	2
E	6	14	−8	8	5

The sum of the rank scores of the positive differences in the Difference column is $1 + 2 = 3$ and the mean rank score of the positive differences is consequently $3/2 = 1.5$.
The sum of the rank scores of the negative differences is $7 + 6 + 8 = 21$ and the mean rank score of the negative differences is consequently $21/3 = 7$.

This indicates that the youngest children are (much) happier than the oldest children.

The greater the number of pairs in which the two members are different in terms of variable Y and, especially, the more frequently that these differences are in the same direction, the smaller is the chance that a difference in Y between two groups is a result of chance. This is even more the case when the sample is larger. The extent to which the samples differ is converted into a z value.

125

6.2.3.2 When do I use the Wilcoxon Signed-ranks test?

The Wilcoxon Signed-ranks test can be used when:

- two related or paired samples are involved;
- the test variable is measured at least at the ordinal level.

The Wilcoxon Signed-ranks test is often used with small samples (< 25), even when the test variable is measured at the interval or ratio levels, and the use of the t test for two paired samples is an obvious option (see Section 6.3.2). With smaller samples, the Wilcoxon Signed-ranks test is less sensitive to extreme scores than the t test is.

When three or more paired groups are to be compared, it is the Friedman test that can be used (Section 6.2.4).

6.2.3.3 How do I perform a Wilcoxon Signed-ranks test?

In a separate study of twenty families, it was investigated if a correlation exists between the extent to which the husbands, wives and their (oldest) children felt themselves to be happy. The children had to be at least 12 years old to participate in the study. The happiness questionnaire discussed in Chapter 1 (see Figure 1.1) was also administered to them. In the file on the website labelled 'FAMILY', you will find the total happiness scores of the husbands, wives and their children. The research question that we are using as an example asks if a difference exists between the happiness scores of husbands and their wives. Such a case involves related samples; the husbands and wives likely influence each other and are, therefore, 'related sample elements'.

related sample elements

To use the Wilcoxon Signed-ranks test, you need to select the "Analyze" and "Nonparametric Tests" commands in the main menu. Next you select "2 Related Samples":

> Analyze ⊙
  > Nonparametric Tests
    > 2 Related Samples ⊙

The two variables that you will test in pairs can only be individually moved to the field beneath the heading "Test Pair(s) List". Mark the first variable as, for example, *HAPMAN*. This is then placed beside "Variable 1" in the "Current Selections" field.

> HAPMAN ⊙ [Current Selections: Variable 1]

Next, mark the second variable as, for example, *HAPWOMAN*, which is similarly placed beside "Variable 2" in "Current Selections". With the aid of the arrow button ▶ , both variables are then moved to the field under "Test Pair(s) List". *HAPMAN* and *HAPWO(MAN)* are placed one after the other.

> hapwoman ⊙ [Current Selections: Variable 2]
> ▶ [move to the Test Pair(s) List field] ⊙
> OK ⊙

126

See figure 6.10.

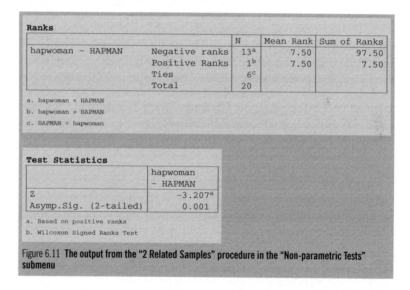

Figure 6.10 **The menu corresponding to "2 related Samples" in the "Non-parametric Tests" submenu**

The output from this procedure is presented in Figure 6.11.

Ranks

		N	Mean Rank	Sum of Ranks
hapwoman – HAPMAN	Negative ranks	13[a]	7.50	97.50
	Positive Ranks	1[b]	7.50	7.50
	Ties	6[c]		
	Total	20		

a. hapwoman < HAPMAN

b. hapwoman > HAPMAN

c. HAPMAN = hapwoman

Test Statistics

	hapwoman – HAPMAN
Z	−3.207[a]
Asymp.Sig. (2-tailed)	0.001

a. Based on positive ranks

b. Wilcoxon Signed Ranks Test

Figure 6.11 **The output from the "2 Related Samples" procedure in the "Non-parametric Tests" submenu**

6.2.3.4 How should I read the results of a Wilcoxon Signed-ranks test?

The output of the Wilcoxon Signed-ranks test consists of two tables: Ranks and Test Statistics. The Ranks table contains the number of 'negative ranks', the number of 'positive ranks' and the number of 'ties'. Of the 20 pairs, 13 involve a negative rank, which is to say a lower score for the women. A higher rank, in which the woman's score is higher than her husband's, only occurs once and in six cases, the happiness scores of the husband and wife are the same.

The test results are presented in the Test Statistics table. Z is –3.21, the corresponding p value is 0.001.
It is difficult to establish the effect size, as the test variable is, in this case, measured at the ordinal level and a conclusion can only be formulated in terms of more or less happy without being able to say anything about how much more and less that it may be.

6.2.3.5 How should I report the results of a Wilcoxon Signed-ranks test?
It is clear that the men have a higher mean rank score on the happiness scale than their wives do. The probability that this difference results from chance is one in a thousand.
In your research report, you could report this result as follows:

"There is a significant difference (Z = –3.21; p < 0.001) in the experience of happiness between husbands and their wives. In the twenty couples included in the study, there were 13 cases involving higher happiness scores for the men. In only 1 case did the wife prove to be happier than her husband. In the remaining couples, there was no difference between husbands and wives in this respect."

Review 6.4
How do I compute the differences in a study involving two paired samples in which there is an ordinal test variable and a nominal grouping variable?

> Analyze ⊙
>> Nonparametric Tests
>> 2 Related Samples ⊙
>>> VAR1A ⊙ [Current Selections: Variable 1]
>>> VAR1B ⊙ [Current Selections: Variable 2]
>>> ▶ [move to the Test Pair(s) List field] ⊙
>>> OK ⊙

6.2.4 Comparative questions involving three or more related samples? Friedman test!

6.2.4.1. What is the Friedman test?
The Friedman test is more or less based on the same principles as the Wilcoxon Signed-ranks test (see Section 6.2.3). The Wilcoxon test is used when there are two paired samples. You use the Friedman test with 3 or more related groups or samples, for example, when you wish to compare the happiness scores of fathers, mothers and their (oldest) children, or when you are dealing with three happiness scores from the same respondents, those collected in year 1, year 2 and year 3. This latter occurs in the

128

context of a so- called panel study. As an example, we will use the happiness scores of the first ten fathers, mothers and their oldest children in the "family" file. For each related group, in this case a family, the scores are ranked from 1 through k (in this case 3).

Table 6.7 The happiness scores of husbands, wives and their children ranked per family

Family	Happiness score			Rank		
	Man	Woman	Child	Man	Woman	Child
1	10	9	10	2.5	1	2.5
2	9	9	10	1.5	1.5	3
3	12	11	12	2.5	1	2.5
4	8	7	9	2	1	3
5	13	12	12	3	1.5	1.5
6	10	10	10	2	2	2
7	12	11	11	3	1.5	1.5
8	9	8	10	2	1	3
9	7	7	8	1.5	1.5	3
10	10	11	12	1	2	3
Total				21	14	23
Mean				2.1	1.4	2.3

For the first family in our data set (Table 6.7), the rank scores are 2.5 for the father, 1 for the mother and 2.5 for the child. In other words, their respective happiness scores are: 10, 9 and 10. As both the father and the child have the highest happiness score, they are ranked between 2 and 3: hence 2.5. And the mother receives the lower rank score of 1; she has the lowest happiness score. Subsequently, the mean rank score is calculated for each group separately (for all the fathers, all the mothers and all the children). Based on the differences between these, that is, instead of the z value (as in the case of the Wilcoxon Signed-ranks test), another value will be used in the conversion into a Chi^2 value.

6.2.4.2 When do I use the Friedman test?
The Friedman test can be used when:
• three or more related samples are involved;
• the test variable is measured at least on the ordinal level.

6.2.4.3 How do I use SPSS to perform a Friedman test?
The 'family' data file involves three groups: fathers, mothers and children. We wish to investigate if there is a difference in the happiness of men, their wives and their children. These three groups can be characterized as related 'samples'. To have a Friedman test performed, you select "Analyze" in the main menu. Then you go to Nonparametric Tests.
> Analyze ⊙
> Nonparametric Tests

In this case, 3 related samples are involved; therefore you choose:
> K Related Samples ⊙

You will then arrive in the "Tests for Several Related Samples" menu (see Figure 6.12). Mark the test variables in the variable list. In this example, they would be *HAPMAN*, *HAPWOMAN* and *HAPCHILD*, all of which are then moved to the field beneath the heading "Test Variable(s)":

> HAPMAN ⊙
> ▶ [to the Test Variable(s) field] ⊙
> HAPWOMAN ⊙
> ▶ [to the Test Variable(s) field] ⊙
> HAPCHILD ⊙
> ▶ [to the Test Variable(s) field] ⊙

[Beeldelement 70]

[fig]Figure 6.12 The menu corresponding to "K Related Samples" in the "Nonparametric Tests" submenu[#fig]

The command can now be executed
> OK ⊙

You can, of course, also select these three variables at the same time by holding down the mouse button as you are making your selection and dragging them all to the "Test Variable(s)" window. The output from this procedure is presented in Figure 6.13.

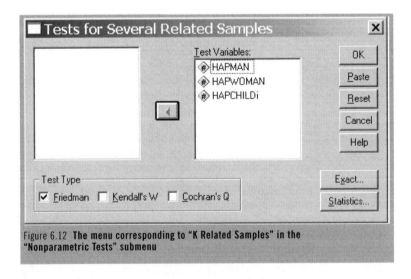

Figure 6.12 **The menu corresponding to "K Related Samples" in the "Nonparametric Tests" submenu**

H6 How do I analyze my data when a comparative research question is involved?

Ranks

	Mean Rank
HAPMAN	2.10
HAPWOMAN	1.27
HAPCHILD	2.63

Test Statistics^a

N	20
Chi-square	23.903
df	2
Asymp. Sig.	0.000

a. Friedman Test

Figure 6.13 **The output resulting from the "K Related Samples" procedure in the "Nonparametric Tests" submenu**

6.2.4.4 How should I read the results of a Friedman test?

The output of the Friedman test consists of two tables: Ranks and Test Statistics.

The Ranks table contains the mean rank scores for the happiness of the different groups. You can see that the women have the lowest scores (1.27). Their husbands clearly score higher (2.10), and their children even higher (2.63).

The test results are presented in the Test Statistics table. Chi² is 23.90, the number of degrees of freedom (df) is 2 (i.e., K – 1, in this example 3 groups – 1). Additionally, twenty ties are shown to be involved. The p value is 0.000.

It is difficult to establish the effect size as the test variable is, in this case, measured at the ordinal level and indicates that you have to draw a conclusion in terms of more or less happy without being able to say anything about how much more and less that it may be.

6.2.4.5 How should I interpret and report the results of a Friedman test?

The differences in the mean rank scores of the *THAP* variable are both substantial and significant. The mothers have the lowest score. The men are clearly happier and the children even more so.

In your research report, you could report this result as follows:

"A comparison of 20 families has shown that wives are clearly less happy (mean rank score 1.27) than their husbands (2.10). The oldest child in each family, in fact, appeared to be the happiest (mean rank 2.63). The differences are significant (Friedman Chi² = 2.90; df = 2; p < 0.001)."

131

Review 6.5

How do I compute the differences in a study involving more than
two related samples in which there is an ordinal test variable and
a nominal grouping variable?

> Analyze ⊙
  > Nonparametric Tests
    > K Related Samples ⊙
    > VAR1A ⊙
    > ▶ [to the Test Variable(s) field] ⊙
    > VAR1B ⊙
    > ▶ [to the Test Variable(s) field] ⊙
    > VAR1C ⊙
    > ▶ [to the Test Variable(s) field] ⊙
  > OK ⊙   ■■■

6.3 Comparative questions involving an interval/ratio test variable and a nominal grouping variable?

6.3.1 Comparative questions involving two independent samples? t test!

6.3.1.1 What is a t test?

t test
The t test is a statistical method of analysis for discovering if the
mean scores of a test variable for two groups (independent samples) differ from each other. The test variable (e.g. *salary*) must
be measured on the interval/ratio level. The grouping variable
(e.g. *sex*) is mostly measured on the nominal level. The probability that this difference results from chance is smaller when:
- the difference between the means of the two samples is larger;
- the variations (the standard deviation) within both samples are
  smaller;
- the samples are larger.

These three elements are included in the formula for t:

$$t = \frac{\overline{X}_1 - \overline{X}_2}{\sqrt{\dfrac{s_1^2}{n_1} + \dfrac{s_2^2}{n_2}}},$$

in which: $\overline{X}_1$ and $\overline{X}_2$ are the respective means from samples 1
and 2

s_1^2 and s_2^2 are the respective variances from samples 1
and 2

n_1 and n_2 are the respective sizes of samples 1 and 2

132

When, for example, the mean IQ in a sample of 50 female employees of the 'Labour' company is 105, while the mean IQ for the sample of 50 male employees is 95, and the standard deviation(s) is 15, then the t value is:

$$t = \frac{105 - 95}{\sqrt{\frac{15^2}{50} + \frac{15^2}{50}}} = 3.33$$

degrees of freedom

The greater the t value the smaller the probability that the difference is a result of chance. The significance of t is certainly dependent on the number of degrees of freedom. In a t test, the number of degrees of freedom is equal to the number of sample elements in sample 1 minus 1 plus the number of sample elements in sample 2 minus 1. In the example, the number of degrees of freedom is therefore $(50 - 1) + (50 - 1) = 98$. For this number of degrees of freedom and a two-tailed testing procedure with an alpha of 0.05, the t value must be at least 2 in order to be significant. Since the computed t value of 3.33 is higher than 2, a significant difference in IQ can be said to exist between the female and male employees of the 'Labour' company.

6.3.1.2 When do I use a t test?

A t test for independent samples can be used when:
- the test variable is normally distributed; this requirement is less important when the sample is larger; but you should nevertheless always check the distribution (Section 5.1.2);
- both samples contain a minimum of 25 respondents;
- the grouping variable is measured at the nominal level; if the grouping variable is measured at the ordinal or interval level, you should ask yourself about the appropriateness of testing if a difference exists between two groups on a given variable; in such a case, you might be better to calculate the correlation between the grouping variable (e.g. *age*) and the test variable (e.g. *income*);
- the standard deviation of the test variables is roughly the same in the two samples.

When two related samples are involved, you will have to use the t test for paired samples (Section 6.3.2).

6.3.1.3 How do I perform a t test for two independent samples?

We will use the 'data5' file. With a t test, you can investigate if the mean age of the men and that of the women are comparable with or different from each other. The comparison of the designated means occurs as follows. You select the "Analyze" command in the main menu and subsequently "Compare Means":

133

6.3 Comparative questions involving an interval/ratio test variable and a nominal grouping variable?

> Analyze ▼
  > Compare Means

Men and women can be viewed as two independent samples. You can then also select the "Independent Samples T Test" command:
  > Independent Samples T Test ▼

Next, you have to indicate the variables that you wish to include in the analysis. The test variable, in this case *age*, is marked and moved in the usual way to the frame beneath the "Test Variable(s)" heading. The grouping variable, in this case *sex*, is moved to the frame under the "Grouping Variable" heading.
  > AGE ▼
  > ▶ [to the Test Variable(s) field] ▼
  > SEX ▼
  > ▶ [to the Grouping Variable field] ▼

Two question marks appear in a position following the grouping variable. Using the "Define Groups" command, you can indicate the two groups that are involved; in our example these are men (code 1) and women (code 2);
  > Define Groups... ▼
    > Group 1 ▼
    > type: 1
    > Group 2 ▼
    > type: 2

If you then click "Continue", you will see that the question marks are replaced by the codes 1 and 2. Clicking "OK" causes the test procedure to be executed.
  > Continue ▼
  > OK ▼

See Figure 6.14.

134

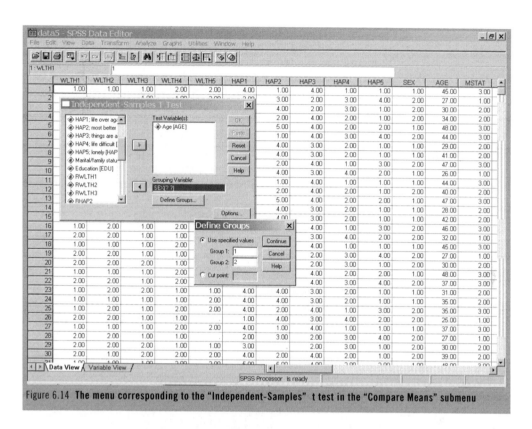

Figure 6.14 **The menu corresponding to the "Independent-Samples" t test in the "Compare Means" submenu**

The output from this procedure is presented in Figure 6.15.

Group Statistics

	Sex	N	Mean	Std. Deviation	Std. Error Mean
Age	Man	498	42.0723	10.06872	0.45119
	Woman	490	37.8102	6.96079	0.31446

Independent Samples Test

		Levene's Test for Equality of Variances		t-test for Equality of Means						95% Confidence Interval of the Difference	
		F	Sig.	t	df	Sig. (2-tailed)	Mean Difference	Std. Error Difference	Lower	Upper
Age	Equal variances assumed	127.612	0.000	7.728	986	0.000	4.2621	0.55154	3.17977	5.34440
	Equal variances not assumed			7.750	884.889	0.000	4.2621	0.54996	3.18271	5.34146

Figure 6.15 **The output from the "Independent Samples" t test in the "Compare Means" submenu**

6.3 Comparative questions involving an interval/ratio test variable and a nominal grouping variable?

6.3.1.4 How should I read the results of a t test for two independent samples?

The output is made up of two tables:

- The 'Group Statistics' table contains, among other things, the number of valid cases (n) for both men and women, the arithmetic mean for the test variable (hence the mean age), the standard deviation and the standard error of mean.

- The 'Independent Samples Test' table is located on the left half of the results for 'Levene's Test for Equality of Variances'. With this procedure, you test the condition that the standard deviation or variance in both samples must be the same, or not significantly different from each other. You may notice that the standard deviations in both samples (10.07 and 6.96) strongly differ from each other. The statistic F, which indicates if the variances (the square of the standard deviation) differ from each other, is therefore in the example rather large, namely 127.61, and hence significant as well (p = 0.000). You cannot, therefore, make use of the test results in the row entitled 'Equal variances assumed'. You have to use the results in the row with the title 'Equal variances not assumed'. In this data, you will see that the t value is 7.75 and the corresponding two-tailed probability of error 0.000. There is, consequently, a significant difference in age between men and women.

It now being established that a significant difference in age between men and women exists, the following step is to investigate the extent to which the differences in age can be explained by sex. This is rather difficult to ascertain on the basis of the t value, since it is related to the number of degrees of freedom. For this reason, the t value is often transformed into Cohen's d in order to determine what the effect size is. The formula appears as follows:

$$d = \frac{2t}{\sqrt{df}}$$

in which: t = t value
df = the number of degrees of freedom
(both are indicated in the SPSS output)

For the example, the formula would read:

$$d = \frac{2 * 7,73}{\sqrt{986}} = 0,49$$

In this case, there is a medium effect. A small effect is one having a d of 0.2, a medium has d = 0.5 and a large d = 0.8. At times, the 'd' is found to be difficult to interpret, as the value is rather specific, insofar as it can vary between 0 and 2. Consequently, the d value is, in fact, converted into an r value, which, simply put, always has a value between 0 and 1 (see the introduction to Chapter 7). In terms of r, 0 indicates that there is no relation and 1 that

136

there is a perfect relation. The transformation into r also makes it possible to calculate the amount of variance in the test variable that can be explained by the grouping variable (see Section 7.3.3). The formula for the transformation is:

$$r = \frac{d}{\sqrt{d^2 + 4}}$$

For the example:

$$r = \frac{0,49}{\sqrt{0,49^2 + 4}} = 0,24$$

This therefore indicates that, of the differences in age, $0.24^2 * 100\% = 6\%$ is explained by sex. Table 6.8 presents a list of d values and the corresponding r and r^2 values.

Table 6.8 **List of Cohen's d and the corresponding r and r^2 values**

Effect size	d	r	r^2
Large	2.0	.71	.50
	1.9	.69	.47
	1.8	.67	.45
	1.7	.65	.42
	1.6	.63	.39
	1.5	.60	.36
	1.4	.57	.33
	1.3	.55	.30
	1.2	.51	.27
	1.1	.48	.23
	1.0	.45	.20
	0.9	.41	.17
	0.8	.37	.14
Medium	0.7	.33	.11
	0.6	.29	.08
	0.5	.24	.06
Small	0.4	.20	.04
	0.3	.15	.02
	0.2	.10	.01
No	0.1	.05	.002
	0.0	.00	.000

6.3 Comparative questions involving an interval/ratio test variable and a nominal grouping variable?

6.3.1.5 How should I interpret and report the results of a t test for two independent samples?

It is clear that a significant difference in mean age exists between men and women. Based on the results of the t test, you could include the following passage in your research report.

"On average, men (mean age 42.07 years old with a standard deviation of 10.07 years) are older than women (mean age 31.81 years old with a standard deviation of 6.96 years). The difference is significant; $t = 7.75$; $df = 884.89$; $p = < 0.001$ in a two-tailed test. Sex can account for 6% of the differences in age ($d = 0.49$)."

Review 6.6

How do I compute the differences in a study involving two independent samples in which there is an interval or ratio test variable and a nominal grouping variable?

> Analyze ▼
>> Compare Means
>>> Independent Samples T Test ▼
>>> VARy ▼
>>> ▶ [to the Test Variable(s) field] ▼
>>> VARx ▼
>>> ▶ [to the Grouping Variable field] ▼
>>> Define Groups... ▼
>>>> Group 1 ▼
>>>> type relevant numerical code
>>>> Group 2 ▼
>>>> type relevant numerical code
>>>> Continue ▼
>> OK

6.3.2 Comparative questions involving two paired samples? t test for two paired samples!

6.3.2.1 What is a t test for two paired samples?

t test for two paired samples

In a t test for two paired samples, the difference is first calculated for each pair. We will use, as an example, the differences in happiness scores between the oldest and the youngest children in a family (see Table 6.9). In this case, the youngest children almost always have higher happiness scores than their older brothers or sisters. The question is whether or not this difference is significant. To determine this, we calculate a t value for paired samples.

138

Table 6.9 Happiness scores of the oldest and the youngest children in the same family			
Pair	Happiness score of oldest child	Happiness score of youngest child	D Difference
1	9	10	−1
2	9	10	−1
3	11	12	−1
4	7	9	−2
5	12	12	0
..	..	..	..
30	12	13	−1
			Mean d = −1.1
			s_d = 0.6

The t value is determined by the size of the differences expressed as the mean difference $(\bar{d})$, the standard deviation of the differences (s_d) and the number of pairs included in the study (n). These factors are taken into account in the formula below. Note: the d value here is concerned with difference and is therefore something else than Cohen's d, which we discussed in the previous section.

$$t = \frac{\bar{d}}{s_d / \sqrt{n}}$$

For the example, this would be:

$$t = \frac{-1,1}{0,6 / \sqrt{30}} = -10,04$$

degrees of freedom

The issue of the t value's significance is decided on the basis of the number of degrees of freedom, which equals the number of pairs minus 1 (df = n − 1). In the example this is, consequently, 30 − 1 = 29. For 29 degrees of freedom and a two-tailed test with = 0.05, the t value has to be at least 2.05 in order to be significant. The computed t value of 10.04 is substantially higher, and a significant difference in the happiness of the youngest and the oldest children in the same family can, therefore, be said to exist.

6.3.2.2 When do I use a t test for two paired samples?
The t test for two paired samples can be used when:
• two dependent or paired samples are involved;
• the test variable is normally distributed; this requirement is less important when the sample is larger; but you should nevertheless always check the distribution (Section 5.1.2); in the case of a skewed distribution and/or extreme scores, it would be more reasonable to use the Wilcoxon Signed-ranks test (Section 6.2.3); **139**

- the sample contains at least 30 pairs of respondents. With a smaller sample, it is more reasonable to use the Wilcoxon Signed-ranks test (Section 6.2.3);
- the grouping variable is measured at the nominal level; if the grouping variable is measured at the ordinal or interval level, you should ask yourself about the appropriateness of testing for a difference between two groups on a given variable; in such a case, it might be better to calculate the correlation between the grouping variable (e.g. *age*) and the test variable (e.g. *income*);
- the test variable is measured at the interval/ratio level.

When you wish to compare three or more related groups, you have to use a multivariate variance analysis, which we are not going to discuss in this introductory book.

6.3.2.3 How do I perform a t test for two paired samples?

In Section 6.2.3, we explained how you can test differences between two related samples in a nonparametric manner. We did this by referring to an example in which the happiness of husbands, wives and the (oldest) children in 20 families were measured. We will recall this example to illustrate how you can use the t test for two paired samples to make such measurements in a parametric manner. Actually, this procedure is not permissible, as the samples are rather small and happiness is not, strictly speaking, measured at the interval level. In practice, you often, however, see that the t test is used with variables that are, in fact, measured at the ordinal level. In such cases, the samples are larger. Once you have opened the 'gezin.sav' file, you once again select the "Analyze" command in the main menu and subsequently "Compare Means":

> Analyze ▼
  > Compare Means

In this case, two paired or related samples are involved. Hence you select "Paired-Samples T Test"

> Paired Samples T Test ▼

See Figure 6.16.

The two variables that you will test in pairs cannot be moved simultaneously to the field beneath the heading "Paired Variables". The move must be made one at a time. Label the first variable something like *HAPMAN*. This is then placed beside "Variable 1" in the "Current Selections" field.

> HAPMAN ▼ [Current Selections: Variable 1]

(margin note) **t test for two paired samples**

(margin note) **Paired-Samples T Test**

140

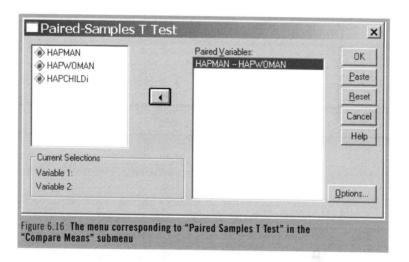

Figure 6.16 **The menu corresponding to "Paired Samples T Test" in the "Compare Means" submenu**

Giving the second variable such a name as *HAPWOMAN*, it is then moved to the same frame beside "Variable 2" : With the aid of the ► button, both variables are then moved to the field under "Paired Variables", where *HAPMAN* and *HAPWO(MAN)* are placed one after the other.

> HAPWOMAN ⊙ [Current Selections: Variable 2]
> ► [to the Paired Variables field] ⊙
> OK ⊙

The output from this procedure is presented in Figure 6.17.

Paired Samples Statistics

		Mean	N	Std. Deviation	Std. Error Mean
Pair 1	HAPMAN	10.0500	20	1.95946	0.43815
	HAPWOMAN	9.4500	20	1.84890	0.41343

Paired Samples Correlations

		N	Correlation	Sig.
Pair 1	HAPMAN & hapwoman	20	0.952	0.000

Paired Samples Test

		Paired Differences							
					95% Confidence Interval of the Difference				Sig.
		Mean	Std. Deviation	Std. Error Mean	Lower	Upper	t	df	(2-tailed)
Pair 1	HAPMAN – HAPWOMAN	0.6000	0.59824	0.13377	0.3200	0.8800	4.485	19	0.000

Figure 6.17 **The output resulting from the "Paired Samples T Test in the "Compare Means" submenu**

6.3 Comparative questions involving an interval/ratio test variable and a nominal grouping variable?

6.3.2.4 How should I read the results of a t test for two paired samples?

The output consists of three tables: Paired Samples Statistics, Paired Samples Correlations and Paired Samples Test. The first-mentioned table contains, among other things, the mean of the test variable for both the husbands and wives (in other words, their mean happiness scores), as well as the standard deviation or dispersion. You will note that the mean happiness score for the men (10.05) is higher than that of the women (9.45). In addition, the 'Paired Samples Correlations' table makes it clear that, whenever the man in a (married) couple feels himself to be very happy, the same holds true for the woman, and vice versa. There is a very strong positive correlation between a husband's experience of happiness and that of his wife. The correlation coefficient is as much as 0.95. In Chapter 7, we will provide more information about correlations. The 'Paired Samples Test' demonstrates that the observed difference in the experience of happiness of husbands and wives is significant. In fact, the t value is 4.49 and the probability that this value is attributable to chance is, for 19 degrees of freedom (the number of respondents minus 1), 0.000. The effect size, that is, the extent to which the differences in happiness scores can be explained by sex, is the mean difference score $(\bar{d})$ divided by the dispersion of the difference scores (s_d). In fact, this should be the mean difference and the standard deviation of the difference scores in the population, but since we do not know what these statistics are, we use the mean difference in the sample and the dispersion of this difference in the sample to make an estimate. These data can be found in the SPSS output:

$$d = \frac{\bar{d}}{s}$$

In the example, the equation would therefore appear:

$$d = \frac{0,60}{0,60} = 1$$

As shown in Table 6.8, there is a large sex-based effect on the happiness score. In contrast to the d value in the t test for independent samples, you cannot just transform this d value into an r value.

6.3.2.5 How should I report the results of a t test for two paired samples?

Although there is a strong relationship between the amount of happiness felt by a husband-and-wife couple, the mean happiness score of men is significantly different from the mean happiness score of women. Based on the results of the t test for related samples, you could include the following text in your research report:

"On average, husbands are happier (mean =10.05; sd = 1.96) than their wives (mean = 9.45; sd = 1.96). The difference is sig-

142

nificant (in a two-tailed test for paired samples, t = 4.46; df = 19; p < 0.001) and the effect of sex on the experience of happiness is a large one (d = 1) For if the husband is happier, the wife, in comparison with other women, is also happier. There is a very strong correlation between the experiences of happiness of a husband and his wife. This was (r = 0,95; p < 0.001). Nevertheless, the women as a group, are less hapy on average."

Review 6.7

How do I compute the differences in a study involving two paired samples in which there is an interval/ratio test variable and a nominal grouping variable?

> Analyze ⓥ
>> Compare Means
>>> Paired Samples T Test ⓥ
>>>> VAR1 ⓥ [Current Selections: Variable 1]
>>>> VAR2 ⓥ [Current Selections: Variable 2]
>>>> ▶ [to the Paired Variables field] ⓥ
>>>> OK ⓥ

6.3.3   Comparative questions involving three or more independent samples? One-way ANOVA!

6.3.3.1   What is a one-way analysis of variance (one-way ANOVA)?

one-way analysis of variance

A one-way analysis of variance is a statistical method of analysis for discovering if the mean scores on a test variable for three or more independent samples differ from each other. Such cases involve test variables measured at the interval/ratio level (for example, *age*). The grouping variable is usually measured at the nominal level, or it is a discrete ordinal variable (for example, *marital status* with the categories: single, with partner, and with partner and children).

The most basic form of variance analysis is the one-way ANalysis Of VAriance (one-way ANOVA). We limit ourselves here to an analysis of variance involving one grouping variable, also known

factor

as a factor.

between- group variance

within-group variance

The total variance in the test variable (*age*) is divided into the between-group variance and the within-group variance. The probability that any difference between groups results from chance is smaller when:

- the variance between groups is larger; hence the differences in (mean) age distinguishing singles, people with partners, and people with partners and children are larger;
- the variance within the groups is smaller; hence the age differences within the separate marital status groups are smaller;

143

- the size of the independent samples being compared is larger; hence when the number of people in the three samples (single, with partner, and with partner and children) is larger.

F value
statistic F
within-group variance
between-group variance
Sum of Squares

The F value is calculated to determine if there is a significant difference in the means of the groups. The statistic F represents the ratio between the mean within-group variance and the mean between-group variance expressed as a Sum of Squares (see ANOVA table in Figure 6.20).

$$F = \frac{MS_{Between}}{MS_{Within}}$$

The 'within-groups sum of squares' is the sum of the squared distances between the individual scores and the group's mean. The 'between-groups sum of squares' is the sum of the squared distances from the scores to the general mean, that is, the mean when you consider all groups together. To determine the mean 'within sum of squares' (MS_{Within}), the 'within sum of squares' is divided by the number of samples minus 1 (= df within). To determine the mean 'between sum of squares' ($MS_{Between}$), the 'between sum of squares' is divided by the total number of respondents minus the number of samples (= df between). When the variance between is (much) larger than the variance within, F will then be (much) larger than 1. The differences in age are then mostly the consequence of the differences in marital status.

The significance of the F value is dependent on the chosen level of significance (alpha) and the df values. The probability of significance is greater when the number of respondents is larger and the number of groups is smaller.

6.3.3.2 When do I use a one-way analysis of variance (one-way ANOVA)?

A one-way analysis of variance is indicated when you wish to investigate if the means of more than two groups on a variable measured at the interval/ratio levels differ from each other.
A one-way analysis of variance can be used when:
- the test variable is measured at the interval/ratio level;
- the test variable is normally distributed; this requirement is less important when the sample is larger; but you should nevertheless always check the distribution (Section 5.1.2);
- the k independent samples each contain a minimum of 25 respondents;
- the grouping variable is of a nominal or a discrete ordinal nature, and does not involve too many categories; if the grouping variable is measured at the interval/ratio level, you should ask yourself about the appropriateness of testing for the differences distinguishing k groups on a given variable; in such a case, you might be better to calculate the relationship between

144

the grouping variable (e.g. *age*) and the test variable (e.g. *income*);
- the standard deviations in the samples are about the same.

If you are dealing with small samples and/or skewed distributions, it might be more advisable for you to conduct a Friedman test (Section 6.2.4).

6.3.3.3 How do I perform a one-way analysis of variance (one-way ANOVA)?

We will again use the 'happiness file' ('data5') as an example. In such a case, you might want to discover if significant differences exist in the (mean) ages of singles (1), people with partners (2), and people with partners and children (3). This requires you to execute the following commands:

> Analyze ⊙
  > Compare Means

Such an issue involves one factor, so we can make use of a one-way ANOVA (ANalysis Of VAriance):

One-Way ANOVA
  > One-Way ANOVA ⊙

In the "One-Way ANOVA" dialogue window that then appears, you next have to indicate the variable involved in this analysis. Mark the test variable *age* in the usual manner and move it to the

Dependent List frame under the heading "Dependent List":

  > age ⊙
  > ▶ [to the Dependent List frame] ⊙

If you would like to include other variables in the "Dependent List", SPSS will perform a separate variance analysis for each listed variable.

Mark the grouping variable in the same way and move it to the

Factor frame under the heading "Factor":

  > marital status ⊙
  > ▶ [to the Factor frame] ⊙

The output will only indicate if there are differences between the groups and if these are significant. No precise indication will be made of between which groups the differences exist, although you will, in most cases, want to have this information. Clicking

Descriptive the "Options" button displays a menu containing the "Descriptive" command. By placing a check in the box in front of this option, you will be given such information, including the mean and the standard deviation of each subgroup:

  > Options... ⊙
    > Descriptive ⊙

See Figure 6.18.

145

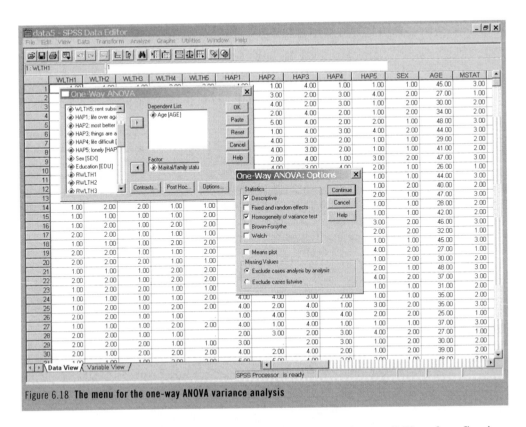

Figure 6.18 **The menu for the one-way ANOVA variance analysis**

The "Options" menu also allows you the possibility of confirming that the condition requiring the variances of the groups to be the same has been satisfied. This confirmation is executed by placing a check in the box beside the "Homogeneity of variance test" command. This action causes the "Levene's Test for Equality of Variances" to be performed (Section 6.4.1):

Homogeneity of variance test
Levene's Test for Equality of Variances

> Homogeneity of variance test ▼
> Continue ▼

Post Hoc

To establish if, and if so, which groups are different from each other, there are several possibilities. If you click the "Post Hoc" button in the "One-Way ANOVA" menu, you will see a list of all the possibilities:

> Post Hoc... ▼

See Figure 6.19.

146

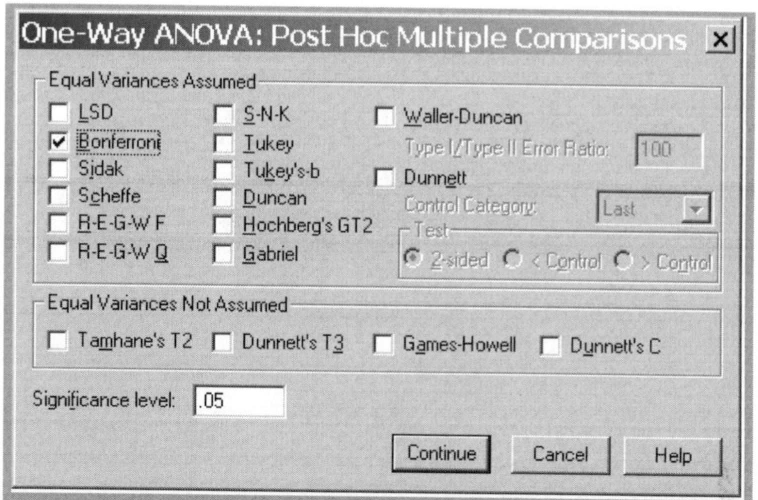

Figure 6.19 **The Post Hoc menu for comparing group means**

In this window, you will also see that you first have to check if the variances are the same. If such is the case, you can then make use of the Bonferroni procedure. This is a simple and transparent technique to investigate post hoc (i.e., after the variance analysis has been executed) which precise groups have significant differences from each other. The procedure is performed as follows:

Bonferroni procedure

> Bonferroni ⊙
> Continue ⊙
> OK ⊙

147

The results from these commands are displayed in Figure 6.20:

Oneway

Descriptives

Age

	N	Mean	Std. Deviation	Std. Error	95% Confidence Interval for Mean		Minimum	Maximum
					Lower Bound	Upper Bound		
alone	150	28.0200	2.94803	0.24071	27.5444	28.4956	25.00	34.00
with partner	365	35.3041	4.59492	0.24051	34.8311	35.7771	28.00	42.00
with partner and children	469	47.4286	5.44099	0.25124	46.9349	47.9223	35.00	55.00
Total	984	39.9726	8.92447	0.28450	39.4143	40.5309	25.00	55.00

Test of Homogeneity of Variances

Age

Levene Statistic	df1	df2	Sig.
37.678	2	981	0.000

Anova

Age

	Sum of Squares	df	Mean Square	F	Sig.
Between Groups	55457.218	2	27728.609	1191.229	0.000
Within Groups	22835.041	981	23.277		
Total	78292.259	983			

Post Hoc Test

Multiple Comparisons

Dependent Variable: Age
Bonferroni

(I)Marial/ family status	(J)Marial/ family status	Mean Difference (I-J)	Std. Error	Sig.	95% Confidence Interval	
					Lower Bound	Upper Bound
alone	with partner	-7.2841*	0.46793	0.000	-8.4062	-6.1620
	with partner and children	-19.4086*	0.45256	0.000	-20.4939	-18.3233
with partner	alone	7.2841*	0.46793	0.000	6.1620	8.4062
	with partner and children	-12.1245*	0.33676	0.000	-12.9320	-11.3169
with partner and children	alone	19.4086*	0.45256	0.000	18.3233	20.4939
	with partner	12.1245*	0.33676	0.000	11.3169	12.9320

*. The mean difference is significant at the .05 level.

Figure 6.20 **The output resulting from the "One-Way ANOVA" in the "Compare Means" submenu**

6.3.3.4 How should I read the results of a one-way analysis of variance (one-way ANOVA)?

We will go over the entire output (see Figure 6.20):

- The 'Descriptives' table contains the means and standard deviations of the various samples. You will notice that the mean age of singles is the lowest and that of people with partners and children is the highest. It is furthermore noteworthy that the standard deviation of age within the singles group is much smaller than in the other groups.

- The 'Test of Homogeneity of Variances' (Levene's test) also demonstrates that the difference in variance amongst the various groups cannot very likely be attributed to chance ($p < 0.001$). In addition, the conditions for the variance analysis are not, in fact, satisfied. This is not a real problem in this example as it involves large samples.

- The 'ANOVA' table indicates that the variance between groups (Mean Square Between: 27728.61) is much larger than the variance within the samples (Mean Square Within: 23,28). The relationship between the 'between variance' and 'within variance', the F value, is large: namely 1190.23. The difference proves to be very significant; the p value is, after all, smaller than 0.001.

- The Bonferroni test in the 'Multiple Comparisons' table indicates that all three groups differ significantly from each other. The probability that the observed differences result from chance is smaller than 0.001. We should, in fact, in this instance, be using a test that does not require us to assume the variances to be equal. For this reason, we have repeated the test using the 'Tamhane's T2 test, in which equivalent variances are not presupposed. This test produces the same result. The probability that the observed differences are a result of chance is for all the comparisons (group 1 with group 2; group 1 with group 3; group 2 with group 3) smaller than 0.001.

Once it has been established that a significant difference exists, it is, of course, also interesting to know the extent to which the difference in age can be explained by marital status. The squared èta (η^2) is often calculated to obtain this information, a statistic that, in fact, expresses the relationship between the differences among the groups and the differences among the totalled scores. η^2 is a measure of the extent to which the variance in a test variable (at the interval level) can be explained by a discrete or nominal grouping variable.

$$\eta^2 = \frac{SS_{Between}}{SS_{Total}}$$

in which: SS = Sum of Squares or the sum of the squared deviation scores

149

In the example, the equation would therefore appear:

$$\eta^2 = \frac{55457}{78292} = 0,71$$

This means that 71% of the differences in age can be explained in terms of the difference in marital status.

6.3.3.5 How should I report the results of a one-way analysis of variance (one-way ANOVA)?

There are significant differences in age distinguishing singles, people with partners, and people with partners and children. You could report this finding in the following manner:

"Based on a one-way analysis of variance, we have observed that significant differences in age distinguish singles (n =150), people with partners (n = 365), and people with partners and children (n = 469)(F(2.981) = 1191.23; p < 0.001). On average, singles are the youngest (mean age 29.02; sd 2.95). People with partners and children are, on average, the oldest (mean age 47.43; sd 5.44). Those living only with a partner occupy an intermediary position (mean age 35.30; sd 4.59). A post-hoc comparison in accordance with the Tamhane method demonstrates that significant differences are shown to exist in all pair-based comparisons (p < 0.001). Seventy-one percent of the differences in age can be explained by marital status ($\eta^2 = 0.71$)."

Review 6.8

How do I compute the differences in a study involving more than two independent samples in which there is an interval or ratio test variable and a nominal grouping variable?

> Analyze ⊙
  > Compare Means
    > One-Way ANOVA ⊙
      > VAR1 ⊙
      > ▶ [to the Dependent List frame] ⊙
      > VAR2t ⊙
      > ▶ [to the Factor frame] ⊙
      > Options... ⊙
        > Descriptive ⊙
        > Homogeneity of variance test ⊙
        > Continue ⊙
      > Post Hoc... ⊙
      > Bonferroni ⊙
      and/or
      > Tamhane's T2 ⊙
      > Continue ⊙
      OK ⊙

H6 How do I analyze my data when a comparative research question is involved?

Key words

151

7

How do I analyze my data when a correlation research question is involved?

> *Remember, time is money*

Benjamin Franklin

1 How do I prepare myself to work with this Introduction?	**2** How do I enter my data in the computer?	**3** How do I modify or combine data?	**4** How do I verify the homogeneity of the composite scores?	**5** How do I analyse my data when a frequency research question is involved?	**6** How do I analyse my data when a comparative research question is involved?	**7** How do I analyse my data when a correlation research question is involved?

7.1 Association between two nominal variables? Cramér's V!	**7.2** Correlation research questions involving ordinal variables? Spearman's rank correlation!	**7.3** Correlation questions involving interval and ratio variables?

We speak of a positive correlation between two variables when a high score on one variable corresponds to a high score on another, and a low score on one is accompanied by a low score on the other. As an example, we refer you to the salary data for 474 bank employees contained in the Employee data file in the SPSS directory (C:\Program Files\SPSS), and to the issue of which factor predominantly determines current salary. Is this the number of months that someone has been in service, someone's educational level and/or a person's starting salary? Such a question concerns the relationship between current salary on the one hand, and service time, educational level or starting salary on the other. You can have SPSS represent such relationships graphically in the form of a scatter plot, commonly known as a scattergram; see Figure 7.1.

Figure 7.1 shows that a positive correlation exists between current salary and beginning salary. An employee with a low starting salary will also have a relatively low current salary, and one having initially earned more will now be earning a relatively large amount. Anyone having begun with 20,000 will have an average current salary of 40,000 and anyone who started by earning 60,000 will, on average, be earning 115,000. This diagram is intended to establish the most suitable, straight regression line, which makes it possible for you to predict such items as someone's happiness score based on that person's wealth score. You can even convert a regression line into an equation. The one corresponding to the line for predicting current salary = 1928 + 1.91 * beginning salary. In Section 2.1, we explain how you can derive such a formula. With a beginning salary of 20,000, you could therefore predict a current salary of 40,128. This agrees nicely with the third graph shown in Figure 7.1. When, in this graph, you trace a beginning salary of 20,000 up to the sloping regression line, you find that a horizontal line

154   extending from a current salary value of 40,000 intersects this

Prior knowledge

We assume that:
- you can start SPSS for Windows (1.4);
- you can retrieve a data file (2.6);
- you can distinguish among the various levels of measurement (1.2.2);
- you know what is meant by the probability of error (p value) and degrees of freedom (= df) (1.4);
- you know when we are talking about significance (1.3);
- you can use SPSS to recode (3.1);
- you know the difference between one and two-tailed tests (1.7);
- you know what a normal distribution is (1.4);
- you know what standard error is (1.4);
- you know what homogeneity entails (4.1);
- you know what expected cell counts are (6.1);
- you can construct a contingency table and calculate the corresponding Chi–square (6.1).

point. The prediction is, however, not completely accurate. In
reality, there are people who had a beginning salary of 20,000
but now earn more than 40,000. The prediction is therefore
imperfect. This is due to our measurements, which are not
entirely situated along one line. Therefore, the data does not
completely fit the line.

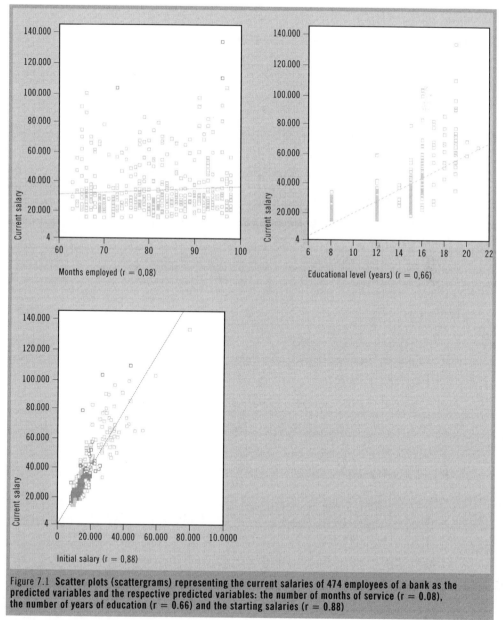

Figure 7.1 **Scatter plots (scattergrams) representing the current salaries of 474 employees of a bank as the predicted variables and the respective predicted variables: the number of months of service (r = 0.08), the number of years of education (r = 0.66) and the starting salaries (r = 0.88)**

Introduction

The extent to which two variables are related and to which one score is, consequently, predictable by the other, is called a correlation. The correlation coefficient therefore indicates the extent that the regression line fits the plot.

When the fit is perfect and you can therefore make perfectly accurate predictions, the correlation is 1. When the line does not conform to the plot at all and you cannot therefore predict anything, the correlation is 0. In the example, the correlation coefficient for beginning salary is 0.88, which means that you can make strong predictions, but any given prediction is not entirely trustworthy. For educational level (Figure 7.1, second graph), the correlation with current salary is 0.66. A prediction is, in this case, even less reliable. This graph also shows that, in the case of educational level, the actual data may lie somewhat further away from the predicted values indicated by the regression line. The number of months that someone is in service (the first diagram in Figure 7.1) has almost no predictive value at all, as the correlation is, in this case, 0.08. This lack of any relationship between the two variables is illustrated by the scatter plot. The measured points lie far away from the predicted ones.

It is possible that, instead of a straight line, a curved one is best suited to this graph. Such a line can also be translated into a regression equation. Given the introductory nature of this book, it is beyond our scope to discuss this subject here. We will limit ourselves here to a discussion of the linear regression.

Besides positive, we also encounter negative correlations. In the SPSS directory, you will find data from an American study comparing all types of cars. Among other things, the vehicle weight and gas mileage of 398 cars are compared, and a negative relationship between these factors clearly demonstrated. The heavier the car, the fewer are the miles that the car can be driven on a gallon of gas. If, in this case, you

request a scatter plot with a linear regression, you will not see

H7 How do I analyze my data when a correlation research question is involved?

an inclining but, in fact, a declining line (Figure 7.2). The
correlation coefficient corresponding to this regression is –0.81.
The fact that the correlation is negative indicates that a negative
relationship is involved.

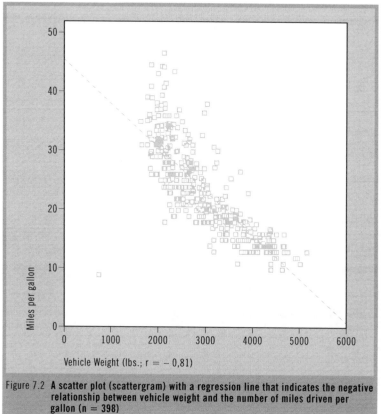

Figure 7.2 **A scatter plot (scattergram) with a regression line that indicates the negative relationship between vehicle weight and the number of miles driven per gallon (n = 398)**

How you calculate the correlation is dependent on the
measurement level of your variables:
- *When the study involves nominal variables, such as sex and*
  marital status, you can neither calculate any correlation nor
  compose a scatter plot. You can still calculate the degree of
  correlation by means of an association measure. It is first
  advisable to construct a contingency table, such as we did in
  Section 6.2. It can be used to examine if a relationship exists
  between such variables as sex and marital status. Section 6.2

157

has, for example, shown that women tend more frequently to live alone than men do. Using an association measure like Cramér's V, you can calculate how strong the correlation is. Cramér's V can also vary from 0, no association, to 1, perfect and therefore completely predictable association. However, no negative associations can, in this case, occur; after all, it is either a higher or a lower relationship that is involved here.

- For measurements at the ordinal level or, when a small sample is involved, at the interval or ratio level, you will have to make use of a form of rank correlation (Section 7.2). An example of a frequently used rank correlation coefficient is Spearman's rank correlation. In this method, the scores are translated into rank scores and used to examine if someone who has a high rank score on one variable also has a high rank score on the other. The stronger that the rank scores agree, the higher the correlation. Viewed in absolute terms, this correlation can also vary from 0 to 1 and can furthermore be negative. When such a negative correlation is involved, a high rank score on one variable is associated with a low score on the other. Since it is only possible to examine if something is higher or lower, and not how much higher or lower it may be, you cannot use this technique to make a regression equation.

- If variables measured on the interval or ratio level are involved and if the sample is not too small (> 30), you can then calculate a Pearson's product moment correlation (Section 7.3). You can also make a regression equation.

For all forms of correlations and associations, it is furthermore important that, whenever you use a sample, you must of course establish if the observed relationship is significant and therefore determine the extent to which the relationship may result from chance. The probability that the observed relationship is attributable to chance not only depends on the strength of the relation, but also on the size of the sample. A Pearson's product

H7 How do I analyze my data when a correlation research question is involved?

moment correlation of 0.30 is, for example, not significant at all

(p > 0.10) for a sample of 10, significant on the 5% level for a

sample of 50 (p < 0.05), and significant on the 1% level for a

sample of 100 (p < 0.01).

7.1   Correlation between two nominal variables? Cramér's V

7.1.1 What is Cramér's V?

contingency table

When we wish to know if there is an association between two variables measured at the nominal level, such as the *sex* and *marital status* variables, it is first prudent to make a contingency table like the one we constructed in Section 6.1.3 (Figure 6.2).
We reproduce the contingency table here in a somewhat more elaborate form (see Figure 7.3).

Marital status * sex Cross tabulation

			Sex		Total
			Man	Woman	
Marital status	single	Count	61	89	150
		% for either sex	12.2%	17.9%	15.1%
	with partner	Count	168	208	376
		% for either sex	33.7%	41.9%	37.8%
	with partner and children	Count	270	200	470
		% for either sex	54.1%	40.2%	47.2%
Total		Count	499	497	996
		% for either sex	100.0%	100.0%	100.0%

Figure 7.3 Contingency table that has been made with "Analyze > Descriptive Statistics > Crosstabs", that has *sex* as the column variable and *marital status* as the row variable, and in which the row percentages have been requested by means of the "Columns" command.

An examination of the contingency table (Figure 7.3) reveals that 18% of the women and 12% of the men are single. There is consequently a small difference and that difference is significant; Chi-square is 19.90 and, when df = 2, p < 0.001 (see Section 6.1). These findings are concerned with differences, and we now wish to investigate the strength of any relationship that may exist. For example, once you know that a person is female, to what extent can you predict that she is single? The table makes it clear that the relationship cannot be a strong one; the difference between the men and the women is, after all, rather small. You can calcu-

Cramer's V

late the association by transforming Chi-square into Cramér's V, which has a value lying between 0, no correlation, and 1, a perfect association.

159

The equation for Cramér's V appears as follows:

$$V = \sqrt{\frac{\chi^2}{N(k-1)}}$$

in which: N = the number of cases; for the example, this consequently amounts to 996

k = the smallest number of columns or rows; for the example, this is 2 (sex)

If you complete the computation, Cramér's V proves to be 0.14; there is hence a very small association between sex and marital status.

phi coefficient

Often, another type of measurement is used: the phi coefficient. It is best employed only when you are using a 2(2 table; however, you then receive the same value as Cramér's V. When a 2(2 table is involved, the formula is therefore the same as the one for Cramér's V. For tables having more than two rows and columns, the phi coefficient can then be greater than 1 and you are, therefore, better off using Cramér's V, as it is easier to interpret.

7.1.2   When do I use Cramér's V?

Cramér's V is used when you wish to know the strength of a relationship between two nominal variables, such as *sex* and *marital status*. First, you should use Chi-square to test if there is a significant deviation from a random distribution (Section 6.1.3).

The requirements for calculating Cramér's V are the same as those for the use of Chi-square, namely:

- any given *expected* cell frequency may not be smaller than 1;
- a minimum of 80% of the *expected* cell-count values are greater than 5;
- the variables must not have too many categories (otherwise the table becomes incomprehensible and the first two conditions are also likely not satisfied).

7.1.3   How do I calculate Cramér's V?

The procedure is more or less the same as the one used to construct a contingency table and to calculate Chi-square (see Section 6.1). Hence:

> Analyze ▼
   > Descriptive Statistics
      > Crosstabs... ▼
         > marital status ▼
         > ► [to the Row(s) field] ▼
         > sex ▼
         > ► [to the Column(s) field] ▼

Clicking "Statistics" opens the "Crosstabs: Statistics" screen (see Figure 7.4):

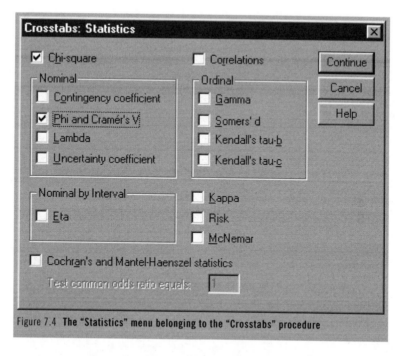

Figure 7.4 **The "Statistics" menu belonging to the "Crosstabs" procedure**

Contingency coefficient
Lambda

In this window, you check both "Chi-square" and "Phi and Cramér's V". You will note that there are still other measures of association, such as the "Contingency coefficient" and "Lambda", but it would lead us too far astray to discuss them here.

> Statistics ⊙
> > Chi-square ⊙
> > Phi and Cramér's V ⊙
> > Continue ⊙
> OK ⊙

Clicking "Continue" and "OK" generates the following output (see Figure 7.5).

Marital status * sex Cross tabulation

Count

		Sex		
		Man	Woman	Total
Marital	single	61	89	150
status	with partner	168	208	376
	with partner and children	270	200	470
Total		499	497	996

Chi-Square Tests

	Value	df	Asymp. Sig. (2-sided)
Pearson Chi-Square	19.904[a]	2	0.000
Likelihood Ratio	19.981	2	0.000
Linear-by-Linear Association	18.309	1	0.000
N of Valid Cases	996		

a. 0 cells (.0%) have expected count less than 5. The minimum expected count is 74.85.

Symmetric Measures

		Value:	Approx. Sig.
Nominal by Nominal	Phi	0.141	0.000
	Cramér's V	0.141	0.000
N of Valid Cases		996	

a. Not assuming the null hypothesis.

b. Using the asymptotic standard error assuming the null hypothesis.

Figure 7.5 **The output from the "Crosstabs" procedure in which requests were made by means of "Statistics" for Chi-square and Cramér's V**

7.1.4   How should I read the Cramér's V output?
Just like the "Crosstabs" procedure discussed above, you are provided with a contingency table (Figure 7.5). The Chi-square test demonstrates that the observed distribution is significantly different from a random one. Specifically, Chi-square is 19.90 and, when df = 2 ((columns − 1) * (rows − 1)), p is < 0.001. The requirement that a maximum of 20% of the expected cell counts are smaller than 5 has also been met, as none of the expected cell counts are smaller than 5 (see Figure 7.5, "Chi-Square Tests", footnote a).

The strength of the association between *sex* and *marital status* is, however, small: Cramér's V is 0.14.

7.1.5   How should I report the Cramér's V output?
The text that you could include in your report strongly resembles the one concerning the Chi-square, only you are now speaking about an association instead of a difference. The passage might appear as follows:

162

"A slight association (Cramér's V - 0.14) can be seen to exist between sex and marital status. Men live together with both a partner and children more frequently than women do. Conversely, women live alone or just with a partner more often than men do. This correlation is significant; Chi^2 = 19.9; df = 2; p < 0.001."

To illustrate this result, you can include the contingency table in your research report (Table 6.2).

Review 7.1

How do I calculate the association between two nominal variables?

> Analyze ⊙
> > Descriptive Statistics
> > > Crosstabs... ⊙
> > > > VAR1 ⊙
> > > > ▶ [to the Row(s) field] ⊙
> > > > VAR2 ⊙
> > > > ▶ [to the Column(s) field] ⊙
> > > > Statistics ⊙
> > > > > Chi-square ⊙
> > > > > Phi and Cramér's V ⊙
> > > > > Continue ⊙
> > > > OK ⊙

7.2 Correlation research questions involving ordinal variables? Spearman's rank correlation!

7.2.1 What is Spearman's rank correlation coefficient (rho)?

In the example study, the research question asks how far wealth makes a person happy. What you would like to know is if a higher score on the wealth scale is accompanied by a higher score on the happiness scale and, conversely, if a lower score on the wealth scale corresponds to a lower score on the happiness scale. In other words, you wish to investigate if a positive relationship exists between these two variables. The items on the happiness scale are measured at the ordinal level. You can, with difficulty, maintain that someone with a total score of 12 is twice as happy as someone with a score of 6. It is, of course, obvious that someone with a score of 12 has a clearly higher happiness score. For the wealth score, things are somewhat more complicated. You might suppose that someone who answered yes four times (Introduction to Chapter 4). has accomplished twice as much as someone with two yes's. Even so, this is not to say that the former respondent has twice as many financial resources at his/her disposal than the latter. The score on the wealth scale can therefore be viewed as an ordinal variable.   **163**

The most commonly used rank correlation is the Spearman's rank correlation coefficient 9(rho). An alternative is Kendall's Tau. We will, however, limit ourselves here to Spearman's rank correlation coefficient.

Spearman's rank correlation coefficient (rho) or r_s is a statistic expressing the extent to which a relationship exists between the ranks of two variables X and Y (for example *wealth* and *happiness*). The scores of the respondents on these variables are therefore initially ranked. Subsequently, the difference in rank on the *wealth* variable and the *happiness* variable can be calculated for each respondent. These differences determine the strength of the correlation.

The value of r_s can vary for –1 to + 1. When r_s = +1, there is then a perfect correlation between both variables. This means that the rank order of variable X is equal to the rank of Y and that no difference in rank order therefore exists. In the case of a completely negative relation, that is, when r_s = –1, the high rank score on the X variable is associated with the lowest rank score on the Y variable, and the lowest rank score on X with the highest Y rank score. In this case, there is a maximum absolute difference in rank scores. When r_s = 0, there is then no correlation and the difference in rank scores lies between 0 and the maximum difference. See Table 7.1.

Table 7.1. **Examples of an entirely positive (r_s = 1.00), a negligible (r_s = 0.10) and a completely negative (r_s = –1.00) rank correlation**

Example r_s = 1.00					Example r_s = 0.10					Example r_s = – 1.00				
Resp	X	RangX	Y	RangY	Resp	X	RangX	Y	RangY	Resp	X	RangX	Y	RangY
A	10	1	50	1	A	10	1	50	1	A	10	1	20	5
B	7	2	40	2	B	7	2	30	4	B	7	2	30	4
C	6	3	35	3	C	6	3	20	5	C	6	3	35	3
D	5	4	30	4	D	5	4	35	3	D	5	4	40	2
E	1	5	20	5	E	1	5	40	2	E	1	5	50	1

When a sample is involved, the significance of the rank correlation is determined by the strength of the correlation and the number of sample elements used to calculate the correlation. For example, an r_s of 0.35 for a one-tailed test at the 5% level is significant when there are 30 sample elements but not significant when there are only 10 sample elements.

7.2.2 When do I use Spearman's rank correlation?

You can compute Spearman's rank correlation when you would like to investigate if a relationship exists between the rank scores of two variables:

- that are measured at the ordinal level;
- or at the interval/ratio level, but when the sample is small (n < 30) and/or contains many extreme scores.

7.2.3   How do I calculate a Spearman's rank correlation?

Suppose that we wish to discover if there is a positive correlation between wealth and happiness. In so doing, we formulate the hypothesis that wealth makes someone happy. In other words, the more financial resources that people have at their disposal, the more content they are about the lives that they lead. We have already indicated that the total scores on both the wealth scale and the happiness scale are ordinal variables.

To calculate the correlation, click "Analyze" in the main menu. We assume that you have retrieved the appropriate data file ('data5') containing the total scores on the wealth and happiness scales. For the sake of absolute clarity and repeating what has already been stated in the homogeneity analysis (Chapter 4), TWLTH is comprised of the sum of the variables RWLTH2, WLTH4 and WLTH5; THAP is equal to the sum of the scores of the variables HAP1, HAP3 and RHAP4. In the "Analyze" roll-down menu, you then select "Correlate" and next "Bivariate", since your analysis does, after all, involve the correlation between two variables:

> Analyze ⊙
> Correlate
> Bivariate ⊙

You must next indicate the variables for which you wish to calculate the correlation. Mark the relevant variables in the usual way and place them in the "Variables" field:

> TWLTH ⊙
> ▶ [to the Variables field] ⊙
> THAP⊙
> ▶ [to the Variables field] ⊙

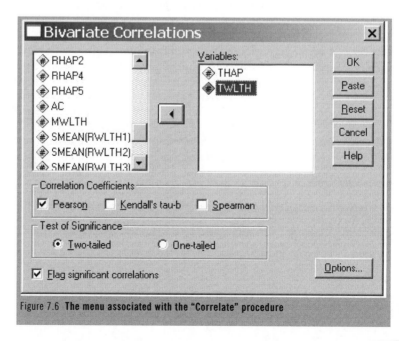

Figure 7.6 **The menu associated with the "Correlate" procedure**   **165**

7.2 Correlation research questions involving ordinal variables? Spearman's rank correlation!

You can then indicate the correlation coefficient to be calculated in the box that has "Correlation Coefficients" as its heading (see Figure 7.6). You can see that, besides Spearman's, you might also choose Kendall's Tau rank correlation coefficient. In this case, we are, however, selecting Spearman's rank correlation.

> Spearman ⓥ

In the box headed "Test of Significance", you need to indicate if the test is to be one or two-tailed. If you already have a certain idea about the direction of the relationship, you should perform a one-tailed test. If you do not have any clear expectation about the direction of the correlation, you will have to conduct a two- tailed test. We have explicitly formulated the hypothesis that a positive correlation exists between wealth and happiness. We will therefore perform a one-tailed test:

> One-tailed ⓥ

> OK ⓥ

Exclude cases pair-wise
Exclude cases list-wise

By clicking the "Options" button, you can then indicate what is to be done with the missing values. The options are: "Exclude cases pair- wise" or "Exclude cases list-wise". If you intend to calculate the mutual correlation between a whole series of variables, a case or respondent with a missing value would be excluded from further consideration in the correlation calculation by 'list-wise deletion'. Sometimes, you then lose a great deal of information, for example, the fact that there are many missing values for one or just a few variables. In this case, it would be better for you to make use of the "Exclude cases pair-wise" option, the default setting for SPSS. However, this does have the disadvantage that the correlations are then calculated separately for different groups of respondents.

By means of "Options", you can also request statistical data about variables for which the correlation is being calculated, such as their means and standard deviations. In Section 5.1.1, we have already discussed a few characteristics of the wealth and happiness scales. For this reason, we need not deal with them here.

By default, the computer places two ** behind correlation coefficients that are significant with an alpha of 0.01 (99% reliability) and one * when you have chosen an alpha of 0.05% (95% reliability). You can also have the ** printed by placing a check in the box next to "Flag significant correlations". The exact p value or probability of error is then also printed.

Finally, click "OK". The output for the procedure can be found in Figure 7.7.

Correlations

Correlations

		THAP	TWLTH
THAP	Pearson Correlation	1	0.749**
	Sig. (1-tailed)	.	0.000
	N	984	902
TWLTH	Pearson Correlation	0.749**	1
	Sig. (1-tailed)	0.000	.
	N	902	908

**. Correlation is significant at the 0.01 level (1-tailed).

Nonparametric Correlations

Correlations

			THAP	TWLTH
Spearman's rho	THAP	Correlation Coefficient	1.000	0.737**
		Sig. (1-tailed)	.	0.000
		N	984	902
	TWLTH	Correlation Coefficient	0.737**	1.000
		Sig. (1-tailed)	0.000	.
		N	902	908

**. Correlation is significant at the 0.01 level (1-tailed).

Figure 7.7 **The output of the "Correlate" procedure in which a one-tailed Spearman test is the choice made in "Correlation Coefficients"**

7.2.4 How should I read the results of a Spearman's rank correlation?

The output consists of the 'Correlations' table, and Spearman's rho is found in the 'Nonparametric Correlations' section of the table. The r_s between wealth and happiness proves to be 0.74 and positive. The p value for this correlation coefficient is 0.000. Moreover, Spearman's rho hardly deviates from the product- moment correlation coefficient (in the "Correlations" table), which in fact amounts to 0.75 in the above example, and which we will discuss below in Section 7.3.2. With larger samples, the two coefficients are, in effect, nearly identical. The correlation is calculated for 902 cases (in this example, respondents).

The p value for the rank correlation coefficient is 0.000. The relationship between wealth and happiness is consequently significant at an alpha of 0.01 and even 0.0001. The p value is, after all, smaller than 0.001. This means that the more wealth that someone has, the happier he or she is. Although the correlation can be said to be rather high, the relationship is not perfect. Happiness can therefore be, to a great extent, explained by wealth, but there are also other factors that have an influence on it. A perfect correlation is almost never found, partly because these sorts of measurements are not completely reliable.

167

7.2.5 How should I report the results of a Spearman's rank correlation?

The following statement can be included in your research report:

"A strong, positive relationship has been found to exist between the financial resources that respondents have at their disposal and the extent to which they are content about the lives they are leading (r_s = 0,74; p < 0,001, one-tailed). The hypothesis that wealthier people are also happier has therefore been confirmed."

Review 7.2

How do I calculate the correlation between two ordinal variables?

> Analyze
> > Correlate
> > > Bivariate ▼
> > > > VAR1 ▼
> > > > ▶ [to the Variables field] ▼
> > > > et cetera through
> > > > VARn ▼
> > > > ▶ [to the Variables field] ▼
> > > > Spearman ▼
> > > > OK ▼

7.3 Correlation questions involving interval and ratio variables?

scatter plot
scattergram

If you assume that a linear correlation between variables exists, it is first advisable to make a scatter plot or scattergram (Section 7.3.1) to see if a correlation is indeed present. If there appears to be one, you can use Pearson's product-moment correlation to calculate how strong the relationship is (Section 7.3.2). If you would like to make predictions, you can then also have SPSS calculate a regression equation (Section 7.3.3). In so doing, you do not have to limit yourself to one predictive variable, but can also involve more than one variable in the prediction (Section 7.3.4). You can, for example, not only predict current salary on the basis of beginning salary, but also on the basis of both beginning salary and the number of years of education.

7.3.1 A scatter plot (scattergram) and a regression line!

7.3.1.1 What is a scatter plot (scattergram) and a regression line?

scatter plot

You use a scatter plot to record the scores of a group of respondents on two variables, for example X and Y. The scale for the X scores is found on the horizontal axis; on the vertical axis you find the scale for the Y scores. If Respondent 1 has, for example, a be-

168

ginning salary (X) of 20,000 and a current salary (Y) of 40,000, you then go along the horizontal X axis to 20,000 and then vertically until you intersect with the horizontal line belonging to 40,000, where you place a point. You do the same, in this case by means of SPSS, for all the respondents. A scatter diagram is created, as shown in Figure 7.1. The more that the scatter diagram resembles a line, the more likely is the case that a linear correlation is involved. You can also request SPSS to draw a straight line, a regression line so that the distances from the points to the line is as small as possible. You need to note the vertical distance to the regression line. The total of the distances from the points to the regression line is called the Residual. It should be evident that the smaller the distances the higher the correlation.

regression line

Residual

7.3.1.2 When do I use a scatter plot (scattergram) and a regression line?

If you presume a linear relationship between two variables measured at the interval and/or ratio level, it is always advisable first to make a scattergram to examine if your presumption is an accurate one. Perhaps there is no relationship; the scatter diagram does not approximate a line in any way. It is also possible that a curvilinear relationship could exist; the scatter diagram does, in fact, involve a relationship represented by a line, but the line is a curve. A well-known example of a curvilinear relationship is the relation between fear of failure and performance. In general, you perform best when you have a slight fear of failure. Your performances are, conversely, somewhat less successful when you either have no fear of failure or a great deal of it.

curvilinear

7.3.1.3 How do I construct a scatter plot (scattergram) and a regression line?

By using the "Scatter" command in the "Graphs" menu, you can construct a scatter plot. The "Graphs" menu can be found on the menu bar at the top of the screen. As you can see, there are various options. Let us just choose the simplest form:

> Graphs
> > Scatter ⊙
> > > Simple ⊙

By means of the "Define" window, you now have to indicate which variable you wish to place on the X axis. We have selected 'Beginning Salary'. 'Current Salary' is accordingly placed on the Y axis.

> > > beginning salary ⊙
> > > ▶ [to the X Axis field] ⊙
> > > current salary ⊙
> > > ▶ [to the Y Axis field] ⊙
> > > OK ⊙

You are then presented with the scattergram illustrated on the right side of Figure 7.1, but without the regression line. The re- **169**

SPSS Chart Editor

gression line can be made to appear by double clicking the scattergram frame in your output. You then arrive in the menu of the "SPSS Chart Editor"; see Figure 7.8. This menu offers you many options for processing your graph; we will discuss them further in Appendix 2. It does, however, provide you with the possibility of drawing a regression line. To do this, you must move the cursor to "Chart" in the menu bar and click it.

You then arrive in the menu of the "SPSS Chart Editor"; see figure 7.8. This menu offers you many options for the processing of your graph; we will examine these further in Appendix 2. However, it also provides the option of drawing a regression line. You first have to click the regression points in the diagram; these are then highlighted. Next, move the mouse cursor to the "Chart" menu option and subsequently to "Add Chart Element", where you select "Fit Line at Total". Instead of these commands, you can also click the icon of the diagram containing a straight line. Choose "Linear". The most suitable straight regression line is drawn in the diagram. You finish the procedure by clicking "Apply" and "Close". By clicking the cross in the top right corner, you return to your output.

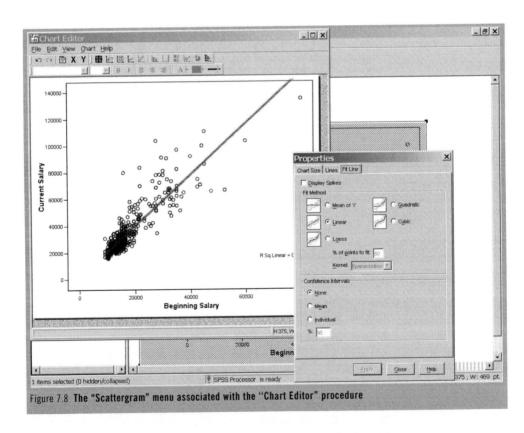

Figure 7.8 **The "Scattergram" menu associated with the "Chart Editor" procedure**

H7 How do I analyze my data when a correlation research question is involved?

> move the mouse cursor to somewhere within the Scattergram
field in your output ⊙
 > Move the cursor to the first variable⊙
 > Chart ⊙
  >Add Chart Element⊙
   > Fit Line at Total ⊙
    > Linear ⊙
    > Apply ⊙
    > Close ⊙
 > X [close] ⊙

Review 7.3
How do I make a simple scatter plot containing a linear regression line?

> Graphs
 > Scatter ⊙
  > Simple ⊙
  > Define ⊙
  > VAR1 ⊙
  > ▶ [to the X Axis field] ⊙
  > VAR2 ⊙
  > ▶ [to the Y Axis field] ⊙
  > OK ⊙
 [add linear regression line]
 > move the mouse cursor to somewhere within the
 Scattergram frame in your output ⊙
  > Chart ⊙
  > Options ⊙
   > Total [Fit Line] ⊙
   > OK ⊙
  > X [close] ⊙

7.3.2   Correlations involving interval and ratio variables?
          Pearson's product-moment correlation!

7.3.2.1   What is a Pearson's product-moment correlation?

Pearson's product-moment correlation coefficient, r

linear relationship

Pearson's correlation coefficient, or more fully, Pearson's product-
moment correlation coefficient, is a statistic which is indicated by
the symbol r. Pearson's r indicates the extent to which a linear relationship exists between the two variables, X and Y, which are
measured at the interval and ratio level. The value of r can range
from -1 to $+1$. When $r = 1$, there is a perfect (linear) positive relationship between both variables. When variable X is high or
low, then the same also holds true for variable Y. In a scatter plot,
all points then lie along an upward-sloping straight line. The
same is also the case with a perfect negative relationship, only
the line is downward-sloping and therefore $r = -1$. A high X cor-   171

responds to a low Y and a low X to a high Y. When r = 0, no linear relationship exists; a(nother type of) relationship between X and Y may, however, still exist.

significance

Whenever a study involves a sample, the significance of the correlation coefficient is dependent on the strength of the correlation and the number of sample elements used to calculate it. A Pearson's product moment correlation of 0.30 is, for example, not significant at all (p > 0.10) for a sample of 10, significant at the 5% level for a sample of 50 (p < 0.05), and significant at the 1% level for a sample of 100 (p < 0.01).

7.3.2.2 When do I calculate a Pearson's product-moment correlation coefficient?

You can calculate Pearson's product-moment correlation coefficient:

linear relationship
curvilinear
relationship

- when there is a linear relationship between two variables (see 7.3.1); when there is a curvilinear relationship, a better alternative may be provided by a measure of association (see Section 7.2). You will then probably first have to reduce both variables to a number of categories by means of the "Recode" command. When you have too many values, you are confronted by an excessive number of cells in your table, which probably also fail to satisfy the condition that a minimum 20% of its cells must have an expected count of at least 5 (see Section 6.1.2);
- when you wish to determine the direction and strength of a correlation between two variables measured at the interval and ratio level;
- when the relationship is of a linear nature. You can check this possibility by means of a scatter plot (see Section 7.4.1);
- when the sample is sufficiently large (> 30) or there is a small sample without any extreme scores.

7.3.2.3 How do I calculate a Pearson's product-moment correlation coefficient?

Broadly speaking, the calculation of the product-moment correlation proceeds in the same way as the rank correlation. As an example, we calculate the product-moment correlation between 'Current Salary' and 'Beginning Salary' to discern how strong the relationship in the scattergram is (Figure 7.8). To do this, you execute the following procedure:

> Analyze ▼
　> Correlate
　　> Bivariate ▼
　　　> Current Salary ▼
　　　> ► [to the Variables field] ▼
　　　> Beginning Salary ▼
　　　> ► [to the Variables field] ▼
　　　> OK ▼

172

The result of this procedure is shown in Figure 7.9.

Correlations

		Current Salary	Beginning Salary
Current Salary	Pearson Correlation	1	0.880**
	Sig. (2-tailed)	.	0.000
	N	474	474
Beginning Salary	Pearson Correlation	0.880**	1
	Sig. (2-tailed)	0.000	.
	N	474	474

**. Correlation is significant at the 0.01 level (2-tailed).

Figure 7.9 **The output of "Pearson's Correlation Coefficient" procedure in the "Bivariate Correlations" menu**

7.3.2.4 How do I read the results of a Pearson's product-moment correlation?

The output (Figure 7.9) is a 'Correlations' table with nine cells, of which there are four that contain three figures each. In the first of these four cells, there is a '1' (the correlation), '.' (the p value) and '474' (n; the number of subjects/elements). This cell involves the correlation of 'Current Salary' with itself, which is, of course, 1. For this reason, the p value is not indicated. The second (next) cell concerns the correlation between 'Current Salary' and 'Beginning Salary'. This correlation is 0.88 and the associated p value is < 0.001, which means that the probability of the correlation resulting from chance is smaller than 0.1%. The correlation is calculated for 474 bank employees. The lower half of the table presents the same data but inverted.

7.3.2.5 How do I make a report on a Pearson's product-moment correlation procedure?

correlation matrix

If there is a correlation, then you record it in your text by including a sentence such as the one below. If it involves the relationship between a number of variables and, consequently, various correlations, it is then useful to construct a table with a correlation matrix (see Section 7.3.4.7).

You could report the relationship between current and beginning salary as follows:

"It has been demonstrated that a relatively strong positive relationship (r = 0.88; p < 0.001; n = 474) exists between current salary and the one earned when hired".

Review 7.4

How do I calculate a product-moment correlation?

> Analyze ▼
  > Correlate
    > Bivariate ▼
      > VAR1 ▼
      > ► [to the Variables field] ▼
      > VAR1 through VARn
      > OK

7.3.3　A simple regression analysis!

7.3.3.1　What is a simple regression analysis?

A linear regression line is, in fact, the graphic representation of a

linear regression formula

linear regression formula. In our example, we would like to predict current salary ($\hat{Y}$) on the basis of beginning salary (X). Such a case involves a prediction at the individual level. For a linear regression, the equation always has the following form: $\hat{Y} = a + bX$, in which:
- $\hat{Y}$ is the variable to be predicted
- X is the variable on the basis of which you wish to predict the value of the other variable

constant
- 'a' is the constant (intercept); it is the point where the regression line intersects the Y axis; this value is calculated with the help of SPSS (see Section 7.3.3.4)

direction coefficient
- 'b' is the direction coefficient (slope) of the regression line; this value is also computed with the help of SPSS (see Section 7.3.3.4).

The regression formula concerning the prediction of the current salary on the basis of the beginning salary is: predicted current salary = 1928 + 1.91 * beginning salary; for the calculation, see Section 7.3.3.4. The figure '1928' is the Y value corresponding to the point where the regression line intersects with the Y axis. When the beginning salary is 0, the predicted current salary would then, according to this formula, be 1928. Now let us briefly consider Figure 7.8. The direction coefficient indicates the direction of the regression line. When this coefficient is positive, the line is an upward sloping one representing, therefore, a positive relationship. When this coefficient is negative, the line is a downward sloping one representing a negative relationship. The example from the introduction (also see Figure 7.2) concerning the relation between the number of miles that a car can drive on a gallon of gasoline and vehicle weight represents a negative correlation. The regression formula is therefore: number of miles on a gallan of gasoline = 45 – 0.007 * weight. Using this formula, you can estimate how many miles that a car can drive per gallon on the basis of the car's weight. In addition, the constant (45) and the direction coefficient (0.007) is derived from the SPSS regression output.

When you work with *standard scores*, in which the mean is 0 and the dispersion 1 (see Section 5.4), the formula becomes somewhat simpler, as follows:

$$z_{\hat{Y}} = b * z_x$$

Note: 'b' has, in this case, another value than the one it has when unstandardized scores are involved. Since the mean for either variable (that is for both $\hat{Y}$ and X) is nil, the constant 'a' is omitted.

174

7.3.3.2 When do I use a simple regression analysis?

When you would like to predict the value of a variable on the basis of another value, you use a linear regression. In that case, you must satisfy the following conditions:

- both variables must at least be at the interval level;
- there must be a strong (r > 0.80) linear relationship (see Section 7.3.2).

7.3.3.3 How do I calculate a simple regression?

SPSS has a special programme for regression. Move the cursor to "Analyze"; there you will find the "Regression" command listed in the roll- down menu. Click it and select "Linear". You are then presented with the menu shown in Figure 7.10.

> Analyze ⊙
>> Regression
>>> Linear ⊙

Figure 7.10 **The "Linear Regression" menu**

Dependent
Independent(s)

In this menu, 'Current Salary' is entered as the to-be-predicted Y in the "Dependent" field. Additionally, 'Beginning Salary' is entered as the predictor X in the "Independent(s) field. You can see that you can enter more than one variable for the predictor X, as it is possible to have several "Independent variables" (we will return to this point in the following section). Finally, clicking "OK" produces the output as shown in Figure 7.11.

> Current Salary ⊙
> ▶ [to the Dependent field] ⊙
> Beginning Salary ⊙
> ▶ [to the Independent(s) field] ⊙
> OK ⊙

175

ANOVA[b]

Model		Sum of Squares	df	Mean Square	F	Sig.
1	Regression	1.07E + 11	1	1.068E + 11	1622.118	0.000[a]
	Residual	3.11E + 10	472	65858997.22		
	Total	1.38E + 11	473			

a. Predictors: (Constant), Beginning Salary
b. Dependent Variable: Current Salary

Model Summary

Model	R	R Square	Adjusted R Square	Std. Error of the Estimate
1	0.880[a]	0.775	0.774	8115.356

a. Predictors: (Constant), Beginning Salary

Coefficients[a]

Model		Unstandardized Coefficients		Standardized Coefficients		
		B	Std. Error	Beta	t	Sig.
1	(Constant)	1928.206	888.680		2.170	0.031
	Beginning Salary	1.909	0.047	0.880	40.276	0.000

a. Dependent Variable: Current Salary

Figure 7.11 **The output from the "Linear Regression" procedure**

7.3.3.4 How do I read the results of a simple regression analysis?

The output of a regression analysis is comprised of four tables, of which the three most important are displayed in Figure 7.11. The most important tables are:

R Square

- Model Summary: This includes the correlation (R = 0.88) and the square of the correlation (R Square = 0.78). The square of the correlation is, in fact, the proportion of the variance in Y (Current Salary) that is explained by X (Beginning Salary). Seventy-eight percent of the differences in current salaries can be explained by a difference in beginning salaries. This means that 22% of the differences cannot be so explained and any prediction will certainly not be perfect. That is also indi-

Std. Error of the Estimate

cated by the 'Std. Error of the Estimate', which amounts to 8115. The estimation error allows you, for example, to indicate the margin within which you can say with 95% certainty what the estimated current salary will be. For instance, the estimated Y will, with 95% certainty, be between $\hat{Y}$ – 1.96 * the estimation error and $\hat{Y}$ + 1.96 * the estimation
error. If, on the basis of the regression formula, the estimated current salary is 100,000, you can therefore with 95% certainty say that the actual salary will lie between 100,000 – (1.96*8115) = 84,094.6 and 100,000 + (1.96*8115) = 115,905.4.

Sum of Squares

- ANOVA: This contains the total amount of variance to be explained (138,000,000,000), the amount explained by the regression formula (107,000,000,000) and the difference between them, the residual. The variance is displayed as the Sum of Squares, which is the sum of the squared differences between

176

the actual and the predicted scores: the distance between the scores and the regression line. The ratio between the amount of variance that has been explained and the total amount of variance to be explained is the square of the correlation (r^2). In this manner, you can also calculate the correlation:

$$r = \sqrt{\frac{explained}{total}} = \sqrt{\frac{107000000000}{138000000000}} = .88$$

Furthermore, this correlation cannot be based on chance because, if you calculate the F value (1622; see Section 6.4.3), you discover that the associated p value is extremely small (p < 0,001).

- 'Coefficients': This is where you find the value that you need for the regression formula. "Constant" represents the constant (intercept), which is 1928 in this case. And 'Beginning Salary' (independent variable) provides you with the direction coefficient (slope), in this case 1.91. The regression formula therefore appears:

estimated current salary = 1928 + 1.91 * beginning salary

When you are working with standardized scores, you also find the 'Beta' here, which is 0.88 for the example. The regression formula for standardized scores therefore appears:

$$z_{estimated\ current\ salary} = 0.88 * z_{beginning\ salary}$$

7.3.3.5 How do I present the results of a simple regression analysis?

The research question must always constitute the basis for your report. In the example, the research question asks if and, if so, to what extent beginning salary is determinative of current salary. You can therefore make the following report:

*"Figure 7.8 demonstrates that there is a clear positive linear relationship between salary granted when hired and current salary. The higher the salary at the start, the higher that it is now, and vice versa: the lower then, the lower now. The strength of the relationship, as expressed by a correlation coefficient, is 0.88 (p < 0.01). The beginning salary can explain 78% of the variance of the current salary. The following formula can be used to predict current salary on the basis of salary received at the start: estimated current salary = (1928 + 1.91) * beginning salary. The corresponding standard error of the estimate is 8115. To be 95% certain of the accuracy of the current salary estimate, a margin must be allowed consisting of the estimated salary plus/minus 1.96 * 8115 = 15,905.4."*

Review 7.5

How do I perform a simple linear regression analysis?

> Analyze ⑦
> > Regression
> > > Linear ⑦
> > > > variable Y ⑦
> > > > ▶ [to the Dependent frame] ⑦
> > > > variable X ⑦
> > > > ▶ [to the Independent(s) field] ⑦
> > > > OK ⑦   ▒▒▒

7.3.4 Correlations involving more than two variables that are measured on the interval or ratio level? Multiple correlations and partial correlations!

7.3.4.1 What is a multiple correlation and regression?

Current salary is probably not only dependent on the salary received when starting. It is advisable to include other characteristics of the bank employees in the study. For the prediction of current salary, we would not only examine beginning salary (X1), but also the work experience at the current job (X2) and the number of months that someone is in service (X3). Y (current salary) can be predicted on the basis of more than one predictor/independent variable. If you would like to know the extent to which a combination of X variables can predict the Y variable, you have to cal-

multiple correlation culate a multiple correlation. In addition, this correlation is based on the extent to which a predicted Y value agrees with the actual Y value. For the prediction, use is made of a regression formula, but you are then using a regression formula with more than one predictive variable. In our case:

estimated current salary = $a + b_1$ * beginning salary + b_2 * work experience + b_3 * months in service

The a and b values are found in the SPSS multiple-regression output.

It should also be noted that, in this case, the formula for standardized scores is somewhat simpler, as it is:

$$z_{estimated\ current\ salary} =$$

$$b_1 * z_{beginning\ salary} + b_2 * z_{work\ experience} + b_3 * z_{months\ in\ service}$$

The Beta values b_1, b_2 and b_3 are indicated in the 'Coefficients' table found in the SPSS multiple-regression output.

7.3.4.2 When do I use a multiple regression analysis?

When you wish to predict one variable on the basis of a number of variables, you employ a multiple linear regression. In that case, you must satisfy the following conditions:

*   the variables must at least be at the interval level;
*   the sample must be sufficiently large (> 30) or a small sample not involving any extreme scores.

7.3.4.3 How do I calculate a multiple correlation?

To calculate a multiple correlation and regression, you actually undertake the same procedures used to make a simple linear regression. You start with "Analyze", then move to "Regression" and then to "Linear". You subsequently arrive in the menu where you enter 'Current Salary' as the "dependent variable" and the others as "independent":

> Analyze ⊙
>> Regression
>>> Linear ⊙
>>>> Current Salary ⊙
>>>> ► [to the Dependent frame] ⊙
>>>> Beginning Salary ⊙
>>>> ► [to the Independent(s) field] ⊙
>>>> Months since Hire ⊙
>>>> ► [to the Independent(s) field] ⊙
>>>> Previous Experience ⊙
>>>> ► [to the Independent(s) field] ⊙
>>>> Method _⊙
>>>> Stepwise⊙
>>>> OK ⊙

See Figure 7.12.

stepwise multiple regression analysis Stepwise

In choosing a "Method", a request is made for a stepwise multiple regression analysis (Stepwise). In each step, SPSS adds each variable to the prediction in the order in which you have entered it. This makes it possible to investigate the extent to which the addition of a given variable improves the prediction. The results are found in Figure 7.13.

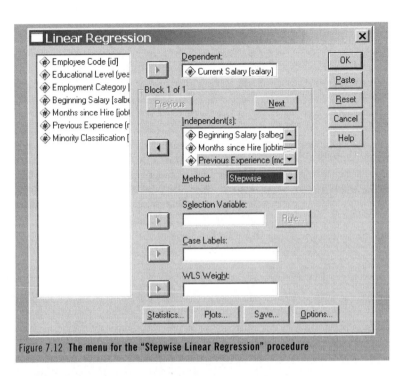

Figure 7.12 **The menu for the "Stepwise Linear Regression" procedure**

Model Summary

Model	R	R Square	Adjusted R Square	Std. Error of the Estimate
1	0.880[a]	0.775	0.774	8115.356
2	0.891[b]	0.793	0.793	7776.652
3	897[c]	0.804	0.803	7586.187

a. Predictors: (Constant), Beginning Salary
b. Predictors: (Constant), Beginning Salary, Previous Experience (months)
c. Predictors: (Constant), Beginning Salary, Previous Experience (months), Months since Hire

Coefficients[a]

Model		Unstandardized Coefficients B	Std. Error	Standardized Coefficients Beta	t	Sig.
1	(Constant)	1928.206	888.680		2.170	0.031
	Beginning Salary	1.909	0.047	0.880	40.276	0.000
2	(Constant)	3850.718	900.633		4.276	0.000
	Beginning Salary	1.923	0.045	0.886	42.283	0.000
	Previous Experience (months)	−22.445	3.422	−0.137	−6.558	0.000
3	(Constant)	−10266.6	2959.838		−3.469	0.001
	Beginning Salary	1.927	0.044	0.888	43.435	0.000
	Previous Experience (months)	−22.509	3.339	−0.138	−6.742	0.000
	Months since Hire	173.203	34.677	0.102	4.995	0.000

a. Dependent Variable: Current Salary

Figure 7.13 **A part of the stepwise multiple regression output**

H7 How do I analyze my data when a correlation research question is involved?

7.3.4.4 How do I read the results of a stepwise multiple regression?

The most important tables are:

- Model Summary: SPSS uses this table to display the multiple correlation for the three predictive models. A prediction based only on beginning salary (1) indicates a correlation of 0.88; if we add previous experience to our considerations (2), the correlation then increases to 0.89; and if we also include the number of months of on-the-job experience (3), the multiple correlation becomes 0.90. These figures make it clear that the predictive value hardly increases when we also consider, in addition to beginning salary, previous experience and the number of months in service. Consequently, it would, in fact, be sufficient for us to base our prediction on beginning salary alone.

- Coefficients: This is where you find the data that you need for the regression formulae; see Section 7.3.3.3. When you wish to make a regression formula for the three predictive X variables, it would then appear as follows:

estimated current salary = − 10.266 + 1.93 * beginning salary − 22.51 * months of previous experience + 173.20 * months in service

When standardized scores are involved, the regression formula appears as follows:

$$z_{estimated\ current\ salary} = 0.89 * z_{beginning\ salary} -$$

$$0.14 * z_{previous\ months\ of\ experience} + 0.10 * z_{months\ in\ service}$$

7.3.4.5 How do I report a multiple correlation and regression?

You could include the following text in your report:

"To predict current salary (Y), we conducted a stepwise multiple regression in which current salary is first predicted on the basis of beginning salary (X_1). The result was a correlation of 0.88 ($p < 0.01$). In step two, previous experience is included in the equation (X_1+X_2); this results in a multiple correlation of 0.89 ($p < 0.01$). The inclusion of the number of months in service as a third variable (X_1+X_2+X_3) yields a multiple correlation of 0.89 ($p < 0.01$). It therefore appears that salary at the beginning could be sufficient for the prediction of current salary. The additional predictive values of previous experience and number of months in service are very limited."

Table 7.1 Summary of the stepwise multiple correlations and regressions of beginning salary (1), experience (2) and number of months in service (3) with current salary as the variable to be predicted ($n = 474$)

Predictive variables	Multiple correlatie with current salary (Y)	Regression equation
X_1 beginning salary X_1 beginning salary	0.88	$Y = 1928 + 1.91X_1$
X_2 experience X_1 beginning salary	0.89	$Y = 3851 + 1.92X_1 - 22.45X_2$
X_2 experience X_3 months in service	0.90	$Y = -10267 + 1.93X_1 - 22.51X_2 + 173.20X_3$

Review 7.6
How do I perform a multiple correlation and regression?

> Analyze ⊙
> > Regression
> > > Linear ⊙
> > > > VARY ⊙
> > > > ▶ [to the Dependent field] ⊙
> > > > VARX1 ⊙
> > > > ▶ [to the Independent(s) field] ⊙
> > > > VARX2 ⊙
> > > > ▶ [to the Independent(s) field] ⊙
> > > > et cetera through VARXn
> > > > Method _ ⊙
> > > > Stepwise⊙
> > > > OK ⊙

7.3.4.6 What is a partial correlation?
When you have various mutually related variables, you can also calculate what the effect is of one variable, when you are controlling for the effect of another. We choose then to conduct another

partial correlation

type of multiple correlation, namely a partial correlation. The relationship between current salary and beginning salary, for example, can also be influenced by educational level. It is likely that beginning salary is, to an important extent, determined by this level. By using a partial correlation, you can calculate what the relationship between two variables is, when you correct for a third variable. In the example, we have calculated the correlation between current salary and beginning salary, while controlling for the number of years of education.

182

7.3.4.7 When do I use a partial correlation?

A partial correlation is used when you would like to investigate if the relation between two variables could be at least partially explained by a third variable. In that case, you must satisfy the following conditions:

*   the variables must at least be at the interval level
*   the sample must be sufficiently large (> 30) or a small sample not involving any extreme scores.

7.3.4.8 How do I calculate a partial correlation?

To calculate a partial correlation, click "Analyze" and "Correlate", but then select "Partial":

> Analyze ⊙
  > Correlate
    > Partial ⊙

You wish to calculate the correlation between current and beginning salary; therefore drag these items into the "Variables" frame:

> Current Salary ⊙
> ► [to the Variables field] ⊙
> Beginning Salary ⊙
> ► [to the Variables field] ⊙

Since you want to correct for the number of years of education, drag this variable into the "Controlling for" frame:

> Educational Level ⊙
> ► [to the Controlling for frame] ⊙

To obtain not only the partial correlations but also the straightforward correlations between the three variables, go to "Options" and click "Zero-order correlations":

> Options ⊙
  > Zero-order correlations ⊙
  > Continue ⊙
> OK ⊙

The results are displayed in figure 7.14.

183

```
- - - PARTIAL CORRELATION COEFFICIENTS - - -

Zero Order Partials

                SALARY      SALBEGIN      EDUC
SALARY         1.0000       0.8801       0.6606
                 (0)         (472)        (472)
                P= ,        P= 0.000     P= 0.000
SALBEGIN       0.8801       1.0000       0.6332
                (472)         (0)         (472)
               P= 0.000     P= ,         P= 0.000
EDUC           0.6606       0.6332       1.0000
                (472)        (472)         (0)
               P= 0.000     P= 0.000     P= ,

(Coefficient / (D.F.) / 2-tailed Significance)
"," is printed if a coefficient cannot be computed
```

```
- - - PARTIAL CORRELATION COEFFICIENTS - - -

Controlling for.. EDUC

                SALARY      SALBEGIN
SALARY         1.0000       0.7948
                 (0)         (471)
                P= ,        P= 0.000
SALBEGIN       0.7948       1.0000
                (471)         (0)
               P= 0.000     P= ,

(Coefficient / (D.F.) / 2-tailed Significance)
"," is printed if a coefficient cannot be computed
```

Figure 7.14 **Output from a partial correlation between 'salary' and 'salbegin' controlling for 'education'**

7.3.4.9 How do I read the results of a partial correlation?

zero order correlations

The output (Figure 7.14) consists of two tables of correlations. The first table contains the straightforward or (zero order correlations). The straightforward correlation between beginning salary and current salary is, for example, 0.88. In the second table, you will find the partial correlation, which is checked for "Educ(ational Level)". This correlation is then 0.79. You will note therefore that a certain portion of the relationship between current and beginning salary is, indeed, explained by the number of years of education. The correlation decreases, after all, only slightly: from 0.88 to 0.79. The number of years of education can therefore only explain a certain amount of the relation, namely: $(0,88^2 - 0,79^2) * 100\% = 17\%$.

7.3.4.10 How do I report a partial correlation?

correlation matrix

It is advisable to mention both the straightforward and the partial correlations. In your report, it is best to record the straightforward (zero-order) correlations in the form of a correlation matrix; see Table 7.2. In so doing, you only report the correlations that are found in the bottom-most triangle of the matrix; if it is done properly, there will only be a series of 1s along the diagonal. This is, after all, the correlation of a variable with itself.

A partial correlation is always designated as $r_{12.3}$, which means that the correlation between variables 1 and 2 has been checked for any effect from variable 3. You can, of course, replace the digits by letters, such as the initial letters of the relevant variables.

The text included in your report could appear as follows:

"It is clear (Table 7.2) that both current salary (r = 0.66); p < 0.001) and beginning salary (r = 0.63; p < 0.001) are correlated with the number of years of education that an employee has acquired. A portion of the relationship between current salary and beginning salary (r = 0.88; p < 0.001) is therefore explained by the number of years of education. Correction for the number of years of education (1) produces, in effect, a lower partial correlation ($r_{12.3}$ = 0,79; p < 0,001) between current salary (1) and beginning salary."

Table 7.2 **The product-moment correlations between current salary, beginning salary and the number of years of education for a sample of 472 bank employees**

1	Current salary	1,00		
2	Initial salary	0,88***	1,00	
3	Years of education	0,66***	0,63***	1,00
		1	2	3

*** significant at the 0.001-level

Review 7.7
How do I perform a partial correlation?

> Analyze ⊙
　> Correlate
　　> Partial ⊙
　　　> X1 ⊙
　　　> ► [to the Variables field] ⊙
　　　> X2 ⊙
　　　> ► [to the Variables field] ⊙
　　　> X3 ⊙
　　　> ► [to the Controlling for field] ⊙
　　　> Options ⊙
　　　　> Zero-order correlations ⊙
　　　　> Continue ⊙
　　　> OK ⊙

186

Appendix 1
How can I include SPSS output in my text?

When you have generated SPSS output, you would probably like to include it in a part of your report. Be nevertheless careful about just copying and pasting extracts from the SPSS output into your text. There are, for example, all sorts of guidelines concerning the format of tables that, for the most part, are not satisfied by the SPSS tables. These latter are furthermore often messy and incomprehensible. Most tables have to be custom made and the figures from the SPSS output used as the contents for them. Table 6.1 is an example of a table constructed according to the rules. In diagrams, such as the "Stem and Leaf Plot" in Figure 5.1, the situation is different; you do not, of course, have to make new versions of them. To insert these in your text, you have to:

a select them
b copy them
c paste them

We will discuss these functions in sequence. To perform these tasks, you will need to have opened both the SPSS programme and a word-processing programme, preferably Word, at the same time.

187

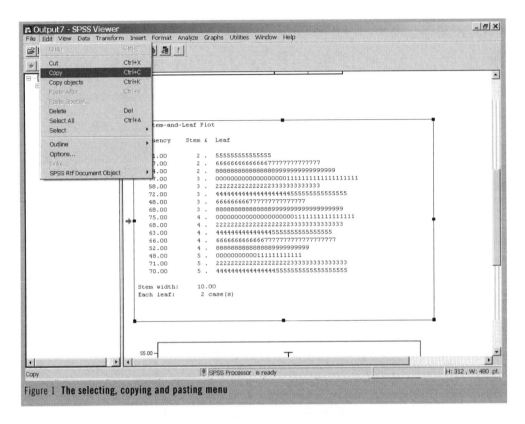

Figure 1 **The selecting, copying and pasting menu**

Edit
Select All

Copy

Ad a: Selecting
Your options include the possibility of selecting everything. You can do this by clicking the Edit roll-down menu and subsequently "Select All". Hence:

> Edit ⊙
> > Select All ⊙

As a result, all output is placed in the blocks.
You can also select parts of it. This is done by clicking the relevant elements. An arrow then appears in front of the selected item (see Figure 1).

Ad b: Copying
You must next indicate that you wish to copy the selected SPSS output. As shown in Figure 1, there are two possibilities: you can select either "Copy" or "Copy objects". If you use "Copy", the selected output will be copied as text. The advantage of this choice is that you can just incorporate the text in a Word document. The disadvantage is that sometimes the output is not as attractively presented as it appears in SPSS. The lines are sometimes lost and the text appears staggered. Because you can process the copied output just as text, it will certainly be easy for you to make any adjustments and to arrange the table to suit yourself.

188

Copy objects

If you select "Copy objects", the SPSS output appears in your text in the original SPSS form. You can no longer alter the copied text. Word recognizes this insert as a chart, one which can be easily reduced or enlarged by merely clicking it. It is then surrounded by a frame. You then move the mouse cursor to a corner of the image. By holding down the left mouse button and dragging the arrow which appears either into or out of the chart, you can give it the desired size. The actual charts contained in SPSS output must be copied as objects without making any alterations.

Ad c: Pasting

Once SPSS output has been selected and copied, it must then be transferred to your text. Move the mouse cursor to the spot where the selected output must appear in your text. Click this spot and then move the cursor to the "Edit" roll- down menu. There you

Paste

will find the "Paste" command. Clicking it will cause the output to be placed at the designated location in your text.

On your screen, you will also see all sorts of icons, such as scissors for cutting next to copies for copying and a clipboard. By clicking the latter, the copied text is placed at the location where the text cursor is located. You can also use key combinations. Ctrl + C is, for example, for the copy command and Ctrl + V for paste.

189

Appendix 2
How can I edit SPSS output in my text?

You can also edit your output in SPSS. This is certainly useful for output that you wish or need to copy as objects, such as 'charts'

Revising graphs

Suppose that you have composed a bar diagram, like the one displayed in Figure 2. Double clicking on the bar diagram causes the

SPSS Chart Editor

SPSS Chart Editor to appear, with which you can execute all sorts of procedures. For example, you can change the colours, as well as the fill patterns in the bars. The latter operation is useful when you wish to make a clear distinction between the data from various

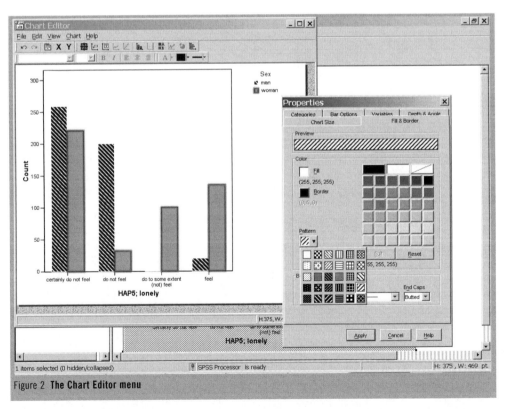

Figure 2 **The Chart Editor menu**

groups, for example, men and women. You then have to click the bar in the diagram displayed on the "SPSS Chart Editor" screen that you wish to change. Then click "Format" on the menu bar, followed by "Fill Patterns", which is where you choose a pattern. By clicking "Apply", you can view the result.

Revising text

In both graphs and other output, you can modify the text by moving the cursor to the text block where you wish to make the modifications. Double clicking this block causes a sort of wavy line to appear around the selected text (see Figure 3). You then again have to double click the text that you wish to change. In the example this is the title, around which a frame composed of straight lines has appeared. It is then possible to alter the text, a revision that we have begun to perform in the example. We have changed the text to 'Contingency table of loneli...'. You see in the Formatting Toolbar that you can also change the font type and size.

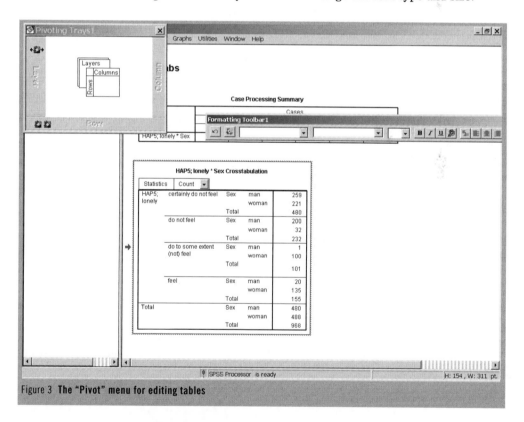

Figure 3 **The "Pivot" menu for editing tables**

Revising tables

In SPSS, there is a special menu for amending tables. First, you again have to double click the table that you wish to change. If you move the cursor to the "Pivot" roll-down menu and click it, you see that you can perform with a click such operations as the   191 rotation of the rows and columns. This could be very useful if you, upon further consideration, would rather have the sex variable in the rows instead of the columns, for example. You can additionally modify the form by using the menu that you see in the "Pivoting Trays" block. To illustrate this point, we have dragged the block in "Column" to "Row", an act that produces a table consisting of rows alone (Figure 3).

Key words

Appendix 1
- edit
- select
- select all
- copy
- copy objects
- paste

Appendix 2
- chart editor
- pivot menu

Appendix 2 How can I edit SPSS output in my text?

Index

193

194

195

196

Serie methoden en technieken

Dr. D.B. Baarda & Dr. M.P.M. de Goede

Een reeks handboeken met aanwijzingen voor het opzetten en uitvoeren van zowel kwantitatief als kwalitatief onderzoek en het rapporteren daarover. Het kenmerk van de boeken is niet alleen het praktijkgerichte karakter, maar ook de daarbij aansluitende didactische opbouw. De hoofdstukken behandelen ieder op zich een aspect van het onderzoekproces. Zij beginnen met de kennis- en vaardigheidsdoelen die bij de betreffende fase van het onderzoek horen en verwijzen naar de begrippen die eerder zijn uitgelegd en die nodig zijn om het hoofdstuk goed te kunnen begrijpen. Elk begrip wordt in een aparte paragraaf uitgelegd en toegepast. Elke paragraaf wordt afgesloten met een samenvatting.

De boeken zijn vooral geschreven voor Hoger Onderwijsstudenten en eerste- en tweedejaars universitaire studenten.

Kwantitatief

Basisboek methoden en technieken
ISBN 90 207 30 30 4

Wat is mijn doelstelling en wat mijn onderzoeksvraag?

Hoe zoek ik informatie?

Wat voor type onderzoek ga ik doen?

Hoe ziet mijn onderzoekspopulatie eruit?

Wat is mijn onderzoekspopulatie en steekproef?

Hoe meet ik mijn begrippen?

Hoe gebruik ik bestaande gegevens?

Hoe ga ik interviewen?

Basisboek enquêteren en gestructureerd interviewen
ISBN 90 207 3086 X

Hoe observeer ik?

Introduction to Statistics with SPSS
ISBN 90 207 3297 8

Hoe prepareer ik mijn gegevens voor de analyse?

Basisboek Statistiek met SPSS voor Windows
ISBN 90 207 3093 2

Hoe rapporteer en evalueer ik?

Kwalitatief

Basisboek Kwalitatief onderzoek
ISBN 90 207 2485 1

Wat is mijn doelstelling en probleemstelling?

Hoe zoek ik informatie?

Welke situatie, groep of persoon kies ik als onderzoekseenheid?

Welke dataverzamelings-methode(n) kies ik?

Participerende observatie, hoe doe ik dat?

Interviewen, hoe doe ik dat?

Basisboek Open Interviewen
ISBN 90 207 2764 8

Bestaande documenten gebruiken, hoe gaat dat?

Hoe registreer en analyseer ik mijn gegevens?

Hoe rapporteer en evalueer ik mijn onderzoek?

Docentenhandleiding; voor docenten per email te bestellen bij voorlichting.ho@wolters.nl onder vermelding van uw naam, privéadres en school waaraan u verbonden bent.